MYSTERIES OF THE Sky

Mysteries of the Sky
Activities for Collaborative Groups

Second Edition – Revised Printing

Shannon Willoughby
Montana State University

Jeffrey P. Adams
Millersville University

Timothy F. Slater
University of Wyoming

Mysteries of the Sky, 2nd Edition – Revised Printing © 2016

First Printing: 1998
Second Printing: 2000
Third Printing: 2013
Fourth Printing: 2016

Front Cover Public Domain Image: Yohkoh SXT

Back Cover Public Domain Image: Yohkoh SXT

ISBN-13: 978-1539408314

ISBN-10: 1539408310

Pono Publishing
PonoPubs.com
604 So. 26th St.
Laramie, WY 82070
USA

Chapters

1. Telescopes — 1
2. The Night Sky — 7
3. Star Magnitudes and Stellar Parallax — 13
4. The Message of Starlight — 19
5. The HR Diagram — 27
6. Stellar Life Cycles — 33
7. Galaxies and the Universe — 41
8. Measuring the Sky — 51
9. Greek Models of the Heavens — 59
10. A Sun-Centered Universe — 67
11. Founders of a Science — 75
12. The Search for the Outer Planets — 83
13. Evolution of the Solar System — 89
14. Space Debris: Asteroids, Meteors, and Comets — 97

Appendices

A. Scales — 105
B. Changing Faces of the Moon — 107
C. The Sun — 111
D. Planetary Data — 115
E. Properties of the Thirty Nearest Stars — 117
F. Properties of the Twenty Brightest Stars — 119
G. Sky Maps — 121

Activities

1. Designing an Observatory — 139
2. Lenses and Telescopes — 147
3. Star Charts — 151
4. Trigonometric Parallax — 153
5. Stellar Spectra Classification — 161
6. HR Diagrams — 168
7. Spectroscopic Parallax — 177
8. Stellar Evolution — 179
9. Galaxy Classification — 181
10. Hubble's Law — 187
11. Distances to the Moon and Sun — 191
12. Tracing Epicycles — 197
13. Determining the Orbit of Mars — 201
14. The Phases of Venus — 203
15. Mapping the Solar System From Earth — 209
16. Moon Phases — 214

Chapter 1

Telescopes

We begin our study of astronomy with a unit on telescopes. Why? Because the telescope is the piece of equipment that is most closely associated with astronomy. Since 1609, when Galileo first began popularizing the use of the telescope for astronomical observations, the telescope has been taking on an increasingly important role in revealing the mysteries of objects in the night sky.

If you look up at the night sky on a dark and clear night you are likely to see as many as 3000 stars. However, even a modest amateur telescope will reveal as many as half a million stars. The telescope shows many more stars because it collects so much more light compared to our eyes, which can intercept only a very small amount of a star's light. For most stars this small amount is insufficient to register in our brains. A telescope's primary job is to gather light and focus it on our eyes, a piece of photographic film, or an electronic detector. This has the effect of amplifying the light making once invisible objects visible. One of Galileo's most profound discoveries was that the Universe contains many more stars than are visible to the naked eye.

Today, advances in technology are providing unprecedented advances in our ability to capture images of distant worlds. Each new generation of astronomers studies a different Universe than that of the previous generation. In fact, now we are able to image entire galaxies of stars so far away that we see them not as they look today but as they looked more than 13 billion years ago. And with each new image providing increasingly detailed information about our Universe, astronomers are struggling to provide consistent scientific theories to help make sense of it all.

1.1 History of the Telescope

As early as 1250 AD, lens makers were using glass to make spectacles (eye glasses) to make small print appear larger to the eyes of aging scholars. Much later, the telescope (called a **spyglass**) was developed and popularized by Dutch lens makers starting in 1608. The principle use of the spyglass was to gain military advantage—it gave one the ability to see flags on distant ships or to count the number of attackers long before they reached the city walls. Italy's Galileo Galilei (1564-1642) was probably not the first person to point a telescope at the heavens, but he is often remembered as the father of telescope astronomy. This is because he vastly improved the technology, appreciated the significance of the wonders he observed, and wrote prolifically about what he saw. Although his telescopes were small and had magnifications ranging only from about 3 to 33 times, he was able to observe craters on the Moon, spots on the Sun, rings around Saturn, Jupiter's moons, and more stars in the Milky Way than had ever been imagined before.

1.2 Three Functions of a Telescope

There are three main functions of a telescope. The first is a telescope's **light gathering power** (LGP). It is the LGP that allows a telescope to gather and focus light from distant and dim objects. The bigger the diameter, or **aperture**, of a telescope, the more LGP it has. Research telescopes, costing millions of dollars and taking years to construct, have very large apertures. Examples of these include the 200 inch Hale telescope on Mt. Palomar near San Diego, the 11 meter SALT telescope in South Africa and the 10 meter Keck telescope on Mauna Kea in Hawaii.

The second most important thing a telescope does is to resolve closely spaced features. This is called **resolving power**. For example, when a car is far away, it is difficult to tell if the car has one headlight or two—you cannot resolve the individual headlights. You have to wait until the car is relatively close to tell if there are two distinct headlights or you are actually seeing a motorcycle with a single light. Our eyes are only a few millimeters in diameter and cannot resolve objects separated by an angle of less than about one arcminute, which is how astronomers express one sixtieth of a degree. For an example of resolving power, consider that as many as half of all the stars you see in the sky that look like single points of light are in fact multiple star systems. But, because the apparent angle of separation between the stars is very small, we tend to see them as single sources of light. We say that we are unable to resolve the two stars because they appear as one. When we talk about the relative positions of stars we are usually referring not to the actual distance between them but rather to their apparent angle of separation as viewed from Earth. The ability to resolve distant objects depends upon the LGP of the telescope.

Finally, the least important (and most heavily advertised of a telescope's functions) is **magnification**. During nearly every tour at an astronomical observatory, visitors ask, "What is the magnification of the largest telescope?" Astronomers generally shudder at this question because magnification is not why astronomers build large telescopes, rather, astronomers use big telescopes because they are often studying very faint objects. To observe a dim object such as a galaxy or a nebula, astronomers need to gather as much light from the object as possible and bring it to a focus. Astronomers who study planets do appreciate magnification and it varies greatly depending on the eyepiece used in the telescope. In general, magnification is the objective focal length divided by the eyepiece focal length. An eyepiece with a smaller focal length gives it a higher magnification, creating a smaller field of view. (Focal length is defined as how strongly a lens focuses light.) The eyepiece is the key to magnification.

Most amateur astronomers probably find that they do most of their observing using a magnification of between 40 and 100 times, which is determined by the choice of eyepiece. Most department stores though are ready to sell you a telescope magnifying 300 to 600 times with an objective of only 2 or 3 inches. These are not recommended and actually, that much magnification is not helpful. The reason is that the size of the objective and the local atmospheric conditions, or "seeing," imposes a much lower limit on the resolution of the telescope. Magnifying more than is required to see the smallest details simply magnifies the distortion. In fact, it often seems that going to a lower power eyepiece actually increases the amount of detail that can be seen because the image is less distorted and brighter—the light is *more* concentrated. The usual rule of thumb for small telescopes is that the maximum useful magnification is about 50 times for each inch of aperture in good seeing conditions. In practice, it can be much less.

1.3   Types of Telescopes – The Refractor

The first spyglasses, or telescopes, were made with two glass lenses. A telescope using only a glass lens for magnification is called a refracting telescope or refractor. The largest lens is at the end of the telescope closest to the object being viewed and it is called the objective lens. The second lens is what the astronomer looks through and it is called the eyepiece lens. Most astronomers agree that refracting telescopes give the clearest and most crisp images. However, there are three main concerns regarding refractors. First, because different colors of light pass through glass lenses differently, stars viewed through a refractor can have circular rainbows around them. Adding a special coating on the objective lens can reduce this chromatic aberration effect. The second concern is weight. Big glass lenses are enormously heavy and difficult to hold. The third concern is expense. Big glass lenses that have no flaws are difficult to make and subsequently expensive.

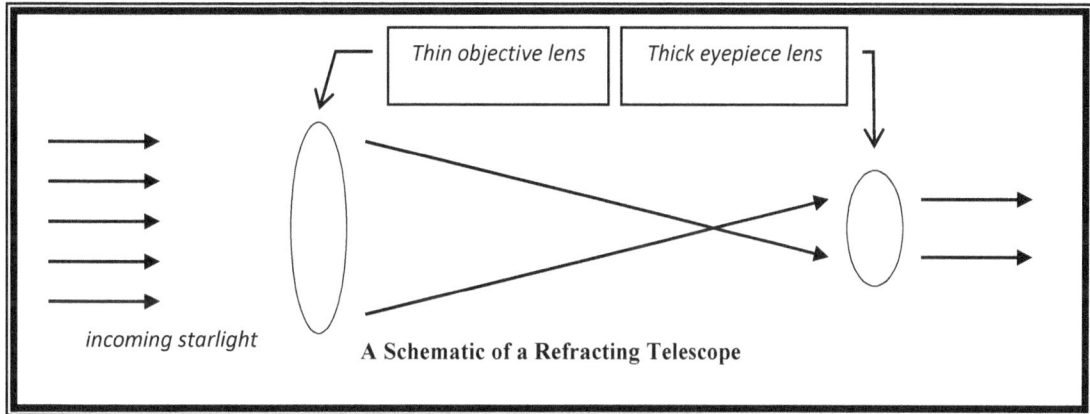

A Schematic of a Refracting Telescope

1.4   Types of Telescopes – The Reflector

A curved mirror is another way to focus a large amount of light. Reflecting telescopes use mirrors instead of lenses. Although the image is not as good as a glass lens, mirrors are lighter, easier to make, easier to hold in place, and much less expensive. The most common reflecting telescope was originally designed by Isaac Newton and is called a Newtonian reflector. The Newtonian reflector uses a curved mirror in the back of the telescope to focus light onto a second, smaller mirror near the front of the telescope. This secondary mirror aims the light to a hole in the side of the tube where the eyepiece is located. The result is that reflecting telescopes are much lighter and much less expensive than refracting telescopes. In fact, the largest telescopes in the world are reflecting telescopes.

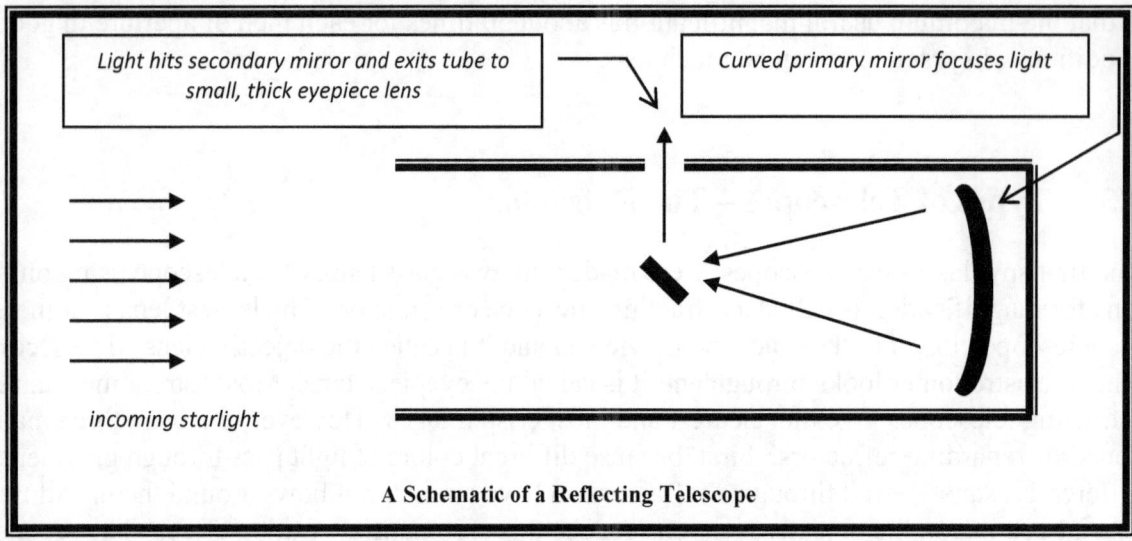

A Schematic of a Reflecting Telescope

1.5 Types of Telescopes – Compound Telescopes

A telescope that takes advantage of the benefits of both glass lenses and mirrors is the compound telescope. Using a lens combined with a mirror makes for a portable telescope that gives fine images for less than the cost of a refractor, but still costs more than a reflector. Many amateur telescopes are of this style. The most commonly used have 8 or 10 inch apertures and weigh about 20 lbs. Two prevalent types are the Maksutov-Cassegrain and Schmidt-Cassegrain.

1.6 Types of Telescopes – Binoculars

The easiest way to start looking at the details of the night sky is with a pair of binoculars. They are cheap, light, and easy to use. Binoculars are described by the following notation: 7 X 35, is read seven-by-thirty-five. The first number is the magnification (7 times in this example) and second number is the diameter of the objective lens in millimeters (35 mm in this example). LGP can be determined based on the diameter. For astronomy, 10 X 50 binoculars work great. However, they will be relatively heavy and astronomers often use a tripod to hold the binoculars still.

1.7 Telescope Mounts

A telescope's pointing and tracking mechanism is crucial to its operation. This point was first appreciated by Galileo, who built large and sturdy mounts for his increasingly powerful and heavy telescopes to ensure that the image did not wiggle. There are several different types of mounts available and the choice of a mount can be as important to the usefulness and enjoyment of a telescope as the choice of optics (lens or mirror).

The Alt-Az Mount

The most basic telescope mount is the altitude-azimuth or "alt-az" mount. An alt-az mount allows the direction of the telescope to be adjusted up-and-down (the altitude) and side-to-side (the azimuth). The simplest alt-az mount is called a Dobsonian, which is normally used to mount a modest (4 to16 inch) Newtonian reflector. The difficulty with this arrangement is that the apparent nightly motion of stars carries them in semi-circular arcs across the sky, which means both the telescope's altitude and azimuth must be continually adjusted to maintain an object in its field of view. It requires some practice to learn how to continually adjust it both up-and-down and side-to-side to follow an object moving across the sky. However, because of the low cost, simplicity, and relative portability of Dobsonian mounted Newtonian reflectors, many consider these to be an ideal first telescope for anyone serious about astronomy.

With the addition of drive motors and a computer guidance system, an alt-az mount can be designed to automatically track the nightly motion of the stars and locate thousands of objects with only the push of a button. As well as making the system easier to use, the ability to automatically track the nightly motion of the sky is critical for most astrophotography. Many higher priced compound telescopes are mounted on alt-az fork mounts using computerized drives for stellar tracking. In fact, the largest telescopes in the world use this type of mount because it can support enormous weight and is easily run when using a newer computer.

The Equatorial Mount

On an alt-az mount, the telescope rotates about an axis that points straight upward. Equatorial mounts, on the other hand, have the axis of rotation aligned to point *straight at the North Star*. This has the distinct advantage that the nightly motion of the stars can be followed simply by rotating the telescope about this single axis with no need to adjust it in the other direction. For manual telescopes, objects can be tracked by adjusting a single knob. More importantly, a simple motor can be used to turn this knob automatically so that once the telescope is pointed at an object it will continue to track without further adjustments. Unlike the alt-az mount, such a system does not require a sophisticated computer system although one certainly can be added.

Even the most inexpensive refractors are generally mounted on equatorial mounts. The most important things to look for in examining such mounts are stability and ease of use. When used at high power, even the smallest jiggle can cause an object to disappear from the field of view; a solid and steady mount is a must. Also, the controls must be easy to use and allow for fine adjustments without having to blindly grope around in the dark to find the correct knob to turn.

Most manufacturers sell a range of Newtonian reflectors either on Dobsonian or equatorial mounts; the optics are the same in both cases. The equatorial mount usually adds between 40 and 100 percent to the cost over the Dobsonian and is decidedly heavier and much less portable. For many, the added cost and inconvenience of the equatorial mount outweighs the benefits.

The Fork Mount

Finally, many compound telescopes use a fork mount in which the fork is tipped sideways on a wedge to create an equatorial mount. These mounts are less stable than the alt-az fork mount but this system provides automatic tracking without the use of a computer. Most systems allow the wedge to be removed if computerized guidance is added to the system.

1.8 Buying Your First Telescope

Deciding to buy your first telescope is an expensive decision. Information and prices can be found in magazines such as *ASTRONOMY* and *SKY & TELESCOPE*. You can also learn a lot by surfing the web. Before you buy you may want to join a nearby amateur astronomy club and attend several meetings. Usually, members have a wide variety of telescopes and will be happy to allow visitors to look through the different models. This will give you a feel for what you like and need. Don't purchase a department store telescope boasting 600 X power. It won't do you much good and will likely have a suspect mounting mechanism. If you have to buy today, go for the 10 X 50 binoculars!

Chapter 2
The Night Sky

From a spot far from the city lights, looking up at the night sky on a clear night can be breathtaking. The grandeur of the sky has provided inspiration for countless poets, painters, and composers. Aside from its innate beauty, the night sky has been historically important for the task of time keeping. Time keeping has always been critical for many aspects of life ranging from planning the harvest to the timing of religious celebrations. The regular changing phases of the Moon and the periodic appearance of different constellations over the year were, at one time, the only way that civilizations had for keeping track of time. The sky-watching task used to be so important that ancient peoples envisioned patterns in the stars and created legends to help generations remember various parts of the sky.

2.1 Constellations and Asterisms

Officially, astronomers divide the night sky into 88 separate regions. These areas have semi-regular shapes, much like the states across the U.S. Each region of the sky is called a **constellation**. Constellations visible from the Northern Hemisphere have Latinized Greek names such as Orion (the hunter), Cygnus (the swan), Ursa Major (the big bear), and Scorpius (the scorpion). Constellations in the southern sky have Latin names, many of which are mariner names such as Telescopium (the telescope), Pyxis (the compass), Horologium (the clock), and Sextans (the sextant). These constellation names and the boundaries between them are coordinated by an international committee of the International Astronomical Union (IAU). The IAU is in charge of official naming issues including naming comets, asteroids, newly discovered moons, and even craters or mountain ranges on planets.

Probably much more familiar to you are the "stick figures" made by drawing lines between the stars. These patterns and recognizable shapes are called **asterisms**. Most people, astronomers included, usually know asterisms better than the official constellations. These include shapes such as the Summer Triangle (between the stars Vega, Deneb, and Altair), the Northern Cross (in the constellation of Cygnus), the Square of Pegasus (in the constellation of Pegasus), and the Big Dipper (in the constellation of Ursa Major). All cultures describe asterisms in their literature and in their legends. Asterisms are completely arbitrary and are used differently among different groups of people to describe locations in the sky.

2.2 Star Names

There are two principal naming conventions for stars. In the Greek System, stars in a constellation are ranked from brightest to dimmest using letters from the Greek alphabet. Accordingly, the brightest star in the constellation of Leo is "alpha Leo," the second brightest star is "beta Leo," and so on. For historical reasons, there are some inconsistencies in the system where, for instance, the second brightest star in a constellation carries the alpha designation. Otherwise, it is a relatively straightforward system.

The second primary naming convention is that the brightest of stars have proper names, mostly Arabic. The brightest star in Leo, alpha Leo, is also known as Regulus (Arabic for King). The bright red star in Orion is Betelgeuse, the bright blue star in Orion is Rigel, and Alpha Scorpius is also known as Antares. This system is simply one of memorization: the star directly above Earth's north pole is known as: (1.) alpha Ursa Minor; (2.) Polaris; and (3.) the North Star.

> **Box 2-1: How Much For That Star?**
>
> There are several companies that purport to sell you a star or name a star after you or a dear friend for a small fee. One such company, the International Star Registry, will assign you your own star, allow you to name it whatever you please, send you a certificate suitable for framing, and provide you with a star map to locate your star (dim enough that you'll likely need a large telescope to find it). Although this might be considered a fun gift, the star registry is not authentic. In fact, if the company wanted to, they could "sell" everyone the same star. All official star names are assigned by an international committee of the International Astronomical Union. Most things astronomical that are currently being named (craters, comets, galaxies) are restricted to being named for their discoverer or to people that have been deceased for some

2.3 Northern and Southern Skies

If you go outside on a clear night and follow the apparent motion of the stars over a period of several hours, you will notice that they all seem to move along curved paths. The only exception is the star Polaris, which is commonly referred to as the North Star. Polaris appears to be fixed in the sky. Stars that are located near Polaris appear to travel counterclockwise along circular tracks centered on Polaris. This includes the familiar stars in the Big Dipper and Cassiopeia. If the path of a star is such that it never dips below the horizon then we call it a **circumpolar star**. These particular stars can be seen on any clear night of the year from that location. It is important to realize however that the stars you see and how they move depend on your viewing location. Depending on where you are located on Earth, you can see different numbers of circumpolar stars. The farther north you are, the more circumpolar stars you can see because the North Star is higher overhead. In fact, at the North Pole, all stars are circumpolar stars. Alternatively, if you are standing at the equator, you cannot see the North Star and no stars are circumpolar. This was a very useful phenomenon for ancient mariners because the altitude of the North Star and the number of circumpolar stars visible allowed navigators to determine their local latitude when far away from shore.

Stars that are located far enough from Polaris that they appear to rise above and then set below the horizon are called **seasonal stars**. These stars rise and set just as our Sun does (for the same reason—a rotating Earth). Even more interesting, if you go every night, you will discover that the stars are in slightly different positions each evening at sunset. Stars rise and set about four minutes earlier on successive nights. Night after night, the constellations appear to be slowly drifting to the west. If a star is seen to rise just at sunset in January—and therefore remain visible all night—it will just be setting at sunset in July and not be visible at all. Of course, the star is still in the sky in July but it cannot be seen because of the overwhelming glow from the daytime Sun. For instance, Orion is apparent in the southern sky during the winter. Six months later, Scorpius is seen where Orion was previously. It is because they can only be seen in certain seasons that we call these seasonal stars. Again, however, this designation depends on the latitude from which you are viewing the sky.

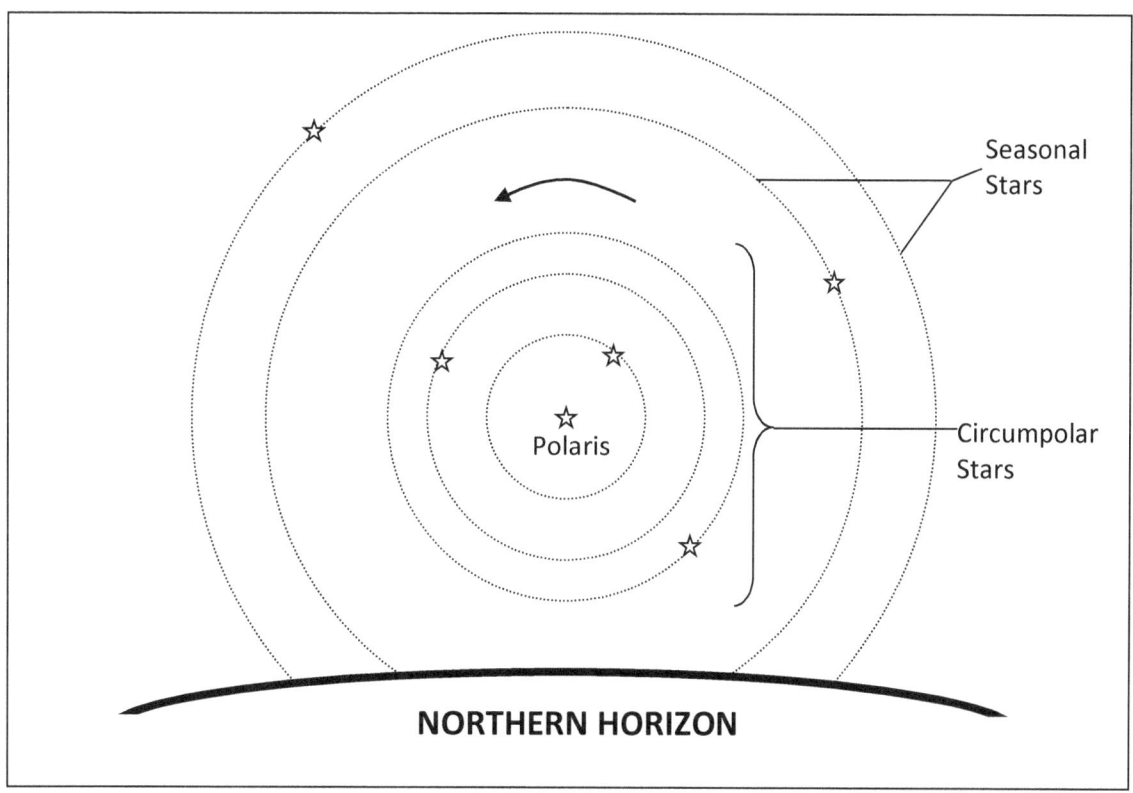

It is this regular and progressive change in the position of the constellations that provides a system of yearly time keeping that is evident in our sky lore. For example, the month of March is known to "come in like a Lion and leave like a lamb." This refers to the fact that at sunset in early March, the constellation of Leo is directly in the southern sky, but, by the end of March, Aires is visible in the evening sky. Another example is the August celebration of "Dog Days." Dog days are the mornings in early August when alpha Canis Major (known also as Sirius and, more commonly known as the Dog Star) is visible in the east just prior to sunrise. It is this regular appearance of seasonal stars and constellations that confused the ancients as to the relationship between time keeping by the stars and the impact of the position of stars and planets on our daily lives.

2.4 Constellations of the Zodiac

As implied earlier, the constellations of the southern sky are only visible in the night sky for about six months. The other six months, these southern sky constellations are in the daytime sky, and are obscured by the brightness of our Sun. As the Sun appears to trace out its annual path relative to the background stars—a path called the **ecliptic**—it actually covers up 13 constellations (it used to be 12, but changes in the Earth's orbit have slightly altered this over the last 2000 years). Most of these constellations are named for animals; these constellations are known more commonly as the constellations of the zodiac (note the prefix zoo as in zoology and as in the place where you go see animals—the zoo). If you have read the local paper, you have probably noticed that folks with different birthdays are assigned a particular constellation in the zodiac as a birth sign. Your assigned birth constellation is the constellation that our Sun is directly in front of on the day you are born; it is not exact though because it is based on the positions from some 2000 years ago and the orbit of the Earth has changed somewhat since. It is, however, fairly close. Keep in mind that the birth signs are slightly different than the zodiacal constellations. For example, Scorpius ⇒ scorpio and Capricornus ⇒ capricorn. Horoscopes are not scientific, as there is no evidence that the location of the planets on your birthdate have any effect on your personality.

Box 2-2: Major Circumpolar and Zodical Constellations Visible from 45° N Latitude

Major Circumpolar Constellations	Constellations of the Zodiac
Ursa Major	Aries
Ursa Minor	Pisces
Cassiopeia	Aquarius
Cephus	Capricornus
Draco	Sagittarius
	Ophiuchus
	Scorpius
	Libra
	Virgo
	Leo
	Cancer
	Gemini
	Taurus

2.5 Using Star Maps

There are twelve monthly star maps found in Appendix G. These, like most others, attempt to plot the expansive and curving overhead sky on a flat piece of paper. No projection method perfectly represents the curved sky on a flat piece of paper and so using a star map takes some practice. To use a star map, there are three things you need to know: (1) what direction you are facing, (2) what time it is, and (3) what month it is.

In general, the first thing to find in the sky and on your map is some prominent star or asterism (the North Star is ideal, as is the Big Dipper, Orion's belt, the Square of Pegasus, or the "j" shape of Scorpius). From there, you can "star hop" to most of the constellations and asterisms. Probably the easiest way to find the North Star is to find the Big Dipper. A line drawn through the last two stars in the cup of the Big Dipper points towards the North Star. The North Star is the last star in the handle of the Little Dipper. Continuing on past the North Star and slightly curving to the right leads you to the "lazy W" of Cassiopeia.

No one learns all of the constellations, or even very many, quickly. Most folks just learn one or two each time they go outside. Astronomers use all sorts of paths, like the one listed above, to find their way around the sky. In the late spring, you can follow the curving handle of the Big Dipper towards the south as, "follow the arc of the handle to Arcturus (*alpha Bootes*) and then spike south to Spica (*alpha Virgo*)." Or you can "draw a line through the belt of Orion down to Sirius (*alpha Canis Major*) or up through Taurus to the seven sisters (*Pleiades*)." It just takes some practice.

Chapter 3
Star Magnitudes & Stellar Parallax

As you look up into the night sky, you will immediately notice that the brightness of stars varies enormously. Some stars are so bright they can be seen before the sky is truly dark while others can only be seen far from the city lights. There are several reasonable explanations for the observed variations in the brightness of stars. One possibility is that all stars have the same light output—called the **luminosity**—and so stars that appear bright are actually close by and stars that appear dim are farther away. Another possibility is that all stars reside at the same distance and some are just more luminous than others. As we will discover later, it turns out that neither hypothesis is quite right—the variations in brightness actually result from differences in both distance and luminosity. Before we deal with this, we first need to understand the system used for describing the brightness and luminosity of the stars.

3.1 Brightness and the Apparent Magnitude Scale

The scale that we use today for describing brightness is called the apparent magnitude scale. It was first established by Hipparchus in about 150 BC. His system was really quite simple. He called the brightest stars in the sky magnitude one; these were stars like Sirius and Vega. Stars like those in the handle of the Big Dipper, the second brightest, were called magnitude two. The third brightest stars were dimmer still, as were fourth, and fifth. Sixth magnitude stars were the dimmest he could see—this was long before the invention of the telescope to see really dim stars. In this scheme, the smaller the magnitude number, the brighter the star.

One way to think about why lower numbers are used for brighter stars is to remember the order in which the stars become visible in the early evening. The first magnitude stars (the brightest) appear first, followed by the second magnitude stars and so on. Thought of as an order of "arrival," the scale makes perfect sense.

Although Hipparchus' scale forms the basis of the modern magnitude system, two important aspects have been changed: negative magnitudes have been introduced and the scale no longer relies on human judgement to measure brightness. Negative magnitudes have been introduced because magnitude 1 stars are not the brightest objects in the sky. This means that we have to use numbers smaller than one for really bright objects. Astronomers now describe Venus as -4, the full Moon as -13, all the way down to -26 for the Sun. Hipparchus' scale was based purely on naked eye observations but astronomers now use electronic detectors to measure the brightness. This has also allowed the scale to be defined more accurately.

The contemporary apparent magnitude scale is actually a measurement of ratios of brightness. Each increase in magnitude of one unit represents a decrease in brightness of about 2.5 times. Thus a first magnitude star is about 2.5 times brighter than a second magnitude star, which is in is about 2.5 times brighter than a third magnitude star. Going farther, what this means is that a first magnitude star is 100 times brighter than a sixth magnitude star. In fact, a difference of 5 in apparent magnitude always represents a difference of 100 times in brightness. This definition also allows us to use fractions for more exact apparent magnitudes. For example, Venus gets as bright as -4.1. To help in your understanding of this, consider the example in Box 3-1.

Even though the Sun appears about ten billion times brighter than Sirius (see Box 3-1), it turns out that Sirius actually emits about 24 times more energy than our Sun emits. The Sun appears ten billion times brighter because it is much closer to us than Sirius. Astronomers actually use two different schemes for describing the **luminosity**, or total energy output, of stars. The first, and most straightforward, is to express the luminosity of a star in terms of the Sun's luminosity, which we will write as L_{sun}. Using this scheme, the luminosity of Sirius is written as $24L_{sun}$, which means that the star Sirius emits 24 times as much light energy as our Sun. The intrinsically much dimmer star Alpha Centauri has a luminosity of only $0.00006L_{sun}$. Even if we were as close to Alpha Centauri as we are to our own Sun, it would still appear very dim indeed. The other scheme that is used for directly comparing the luminosities of stars is the absolute magnitude scale, which is similar to the apparent magnitude scale but with one critical difference.

> **Box 3-1: Magnitude Self Check**
>
> Our Sun has an apparent magnitude of -26. This would make it 100x brighter than a hypothetical star of magnitude -21, 100x100x brighter than a star of magnitude -16, ..., and 100x100x100x100x100 = ten billion times brighter than a star of magnitude -1, which is approximately the magnitude of the star Sirius.

3.2   Luminosity and the Absolute Magnitude Scale

The most convenient way to visually compare the luminosities of stars would be to move them all to some common distance. Then we could say for certain that the stars that appeared the brightest were indeed the most luminous. We cannot actually move stars physically but, in a sense, we can do it mathematically. We do this by calculating how bright stars would appear, expressed in terms of magnitude, if they could all be moved to a common distance 10 parsecs.
*(Astronomers describe the distances to stars using two different units. The **light-year** (ly) is the distance that light travels in one year. The **parsec** (pc) is equal to 3.26 light-years. The origin and definition of the parsec will be discussed later in this unit.)*

For example, if our own Sun were located 10 parsecs from Earth, it would have a magnitude of only 4.8 and be barely

---

Box 3-2: The One Over R Squared Law

We have already alluded to the fact that there is an intimate connection between the brightness of and distance to a star. In fact, the brightness follows a "one over r squared" law. The one over r squared principle is a fundamental concept in physics. The effects of sound, light, and gravity all decrease with increasing distance in the same way, which is described by this law. Consider three stars with identical luminosities located 1, 2, and 3 light-years away. Compared to the star at 1 light-year, the star at 2 light-years appears only 1/4 as bright (i.e., $1/2^2$). The star at 3 light-years appears only 1/9 as bright (i.e., $1/3^2$). That two stars have the same luminosity and yet one appears one quarter as bright as the other does implying that the dimmer star must be located at twice the distance. On the other hand, if two stars appear the same brightness in the sky but we know that one is at three times the distance of the other, then the farther one must have nine times the luminosity.

---

visible to the naked eye. The star Rigel, in contrast, would have a magnitude of –7.2 at 10 parsecs, which would be brighter than any object in the night sky. We can conclude that Rigel is far more luminous than our Sun. These values (4.8 and –7.2) are called the **absolute magnitudes** of the Sun and Rigel respectively and the much lower absolute magnitude of Rigel simply means that it has a much greater luminosity than our Sun. In fact, given the absolute magnitude of a star, its luminosity in units of L_{sun} is relatively easy to compute and visa versa (see box 3-3). The terms absolute magnitude and luminosity are often used interchangeably because, although they use different scales, they refer to the same fundamental property—the total light output of a star.

Box 3-3: The Connection Between Absolute Magnitude and Luminosity

If the Sun were located at a distance of 10 parsecs is would appear as a magnitude 4.8 star. Rigel, located at the same distance would appear as a magnitude –7.2 star. The magnitude of Rigel would be 12 lower than the Sun, which means it would be emitting 2.5x2.5x2.5x2.5x2.5x2.5x2.5x2.5x2.5x2.5x2.5x2.5 times more light energy than the Sun. This number works out to be about 50,000. Therefore, we could also say that the luminosity of Rigel is about $50,000 L_{sun}$.

The one important point that we have not addressed is how these luminosities, expressed as absolute magnitudes, are determined. Only the apparent magnitude can be measured directly. Just by looking at the brightness of a star there is no way to distinguish between a nearby *low* luminosity star from a distant *high* luminosity star. The key to knowing the difference relies on our ability to accurately measure the distance to stars.

3.3 Measuring the Distance to the Nearest Stars by Trigonometric Parallax

Your ability to distinguish the distance to an object—like a clock, a chair, or an oncoming car—is called depth perception. Humans have depth perception primarily because we have two eyes that can work independently. The 2-inch separation between our eyes allows us to look at objects from differing perspectives. Consider holding up your right thumb at arm's length and looking at it with your right eye only. Your thumb will appear at a certain point compared to the background of more distant objects. If you then close your right eye and open your left, your thumb's position relative to the background will mysteriously appear to move slightly to the right. This phenomenon is called parallax—the apparent change in position of an object due to a change in observer perspective. By looking at stars from different observing positions in the Earth's orbit (using a 186 million-mile separation distance) astronomers are able to use parallax to measure the distance to nearby stars.

Box 3-4: Measuring Distances Indirectly with Trigonometric Parallax

Consider trying to determine the width of a river without actually crossing it. The task would be to mark a spot on your side of the river that is exactly across from an obvious object, such as a tree, on the other side. Then, you could walk 100 feet down stream and look back at the tree on the opposite side.

Using a protractor to measure the angle θ and some simple trigonometry yields the distance to the tree. The angle marked p, which is just $90° - \theta$, is called the parallax angle. The smaller the parallax angle, the farther away the object is located.

Because stars are so very far away, the parallax angle p (see figure on opposite page) is very small indeed. It was not until 1837 that astronomers were first able to measure the tiny parallax that could then be used to measure with some accuracy the distance to a nearby star. To measure such small angles, astronomers divide the degree up into 60 units called arcminutes and further divide each arcminute up into 60 arcseconds. Thus, one arcsecond is 1/3600 of one degree. The parallax angle for the nearest stars is on the order of one arcsecond—for more distant stars, p is even smaller.

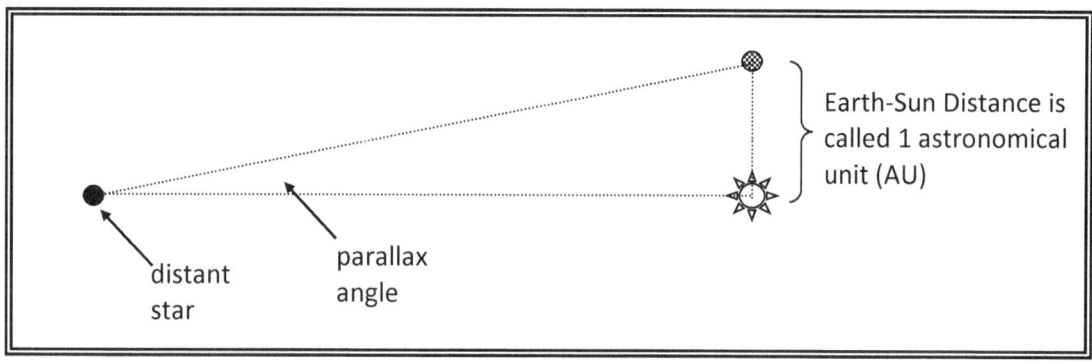

One of the byproducts of using parallax to measure distances to nearby stars is that it offers another unit for expressing the distances to stars. The unit is called the PARSEC, which is a contraction of the words PARallax SECond. The parsec is defined as the distance at which a star must be located to have a parallax angle p of 1 arcsecond. Equivalently, if you were an observer on a star located one parsec from us, the maximum separation angle between the Earth and the Sun would be one arcsecond (1/3600 of a degree). The parsec is a natural unit for describing the distances to stars because it is directly related to the quantity that is actually measured. We must be clear though that the parsec is **not an angle**; more distant stars display smaller parallax angles and yet have distances that are expressed as a greater number of parsecs. For our purposes, what is most important to understand that the parsec is a measure of distance and is equal to about 3.26 ly.

Box 3-5: The Hipparchos Mission

The second-century BC astronomer Hipparcus of Nicea developed what has evolved into the modern magnitude scale and also catalogued the positions of more than 1,000 stars using the only instrument available to him—his eyes. On August 8, 1989 his namesake—the Hipparcos satellite—was launched. Hipparcos, which stands for HIgh Precision PARallax COllecting Satellite, was to spend three and a half years making measurement of the positions and apparent magnitudes of more than one hundred thousand stars with unprecedented accuracy. Exceeding the initial design specifications, Hipparchos was able to measure parallax angles to an accuracy of 1 milliarcsecond (that is one one-thousandth of an arcsecond).

Before Hipparchos, astronomers had calculated the distance to only a few dozen stars with accuracy better than one percent. Hipparchos has increased this number to over 400. At the five percent level, Hipparchos has increased the number of stars from about 400 to over 7,000. However, even with this incredible accuracy, it is important to remember that this still only represents a small fraction of the stars of our own Milky Way Galaxy.

The difficulty that astronomers had in measuring stellar parallax becomes quite clear when you realize that there is no star within one parsec of Earth. The closest is Proxima Centauri, which is located at 1.3 parsecs or about 4.3 ly.

So, how far away can stars be and still have their distances determined using stellar parallax? The limit seems to be about 250 parsecs for very bright stars, which may sound impressive (and as a technological accomplishment it is), but this still encompasses only a very small fraction of the estimated one hundred billion stars in our own galaxy. In fact, the limit of 250 parsecs represents only about one one-hundredth of the estimated diameter of our galaxy. A map of our galaxy requires some other method of measuring distance that does not rely on the motion of the Earth.

3.4 The Method of Standard Candles

We have already seen how knowing how bright a star looks (apparent magnitude) and its distance can be used to determine how bright a star really is (absolute magnitude or luminosity). Since trigonometric parallax only works for relatively nearby stars, we can use the apparent and absolute magnitudes to determine a star's *distance* from us. But to do this, we must know the star's luminosity, or how much light the star is putting out. This method is known as the method of standard candles, because we must use some other method to determine the luminosity of the star in order to determine its distance from us.

Imagine that you are walking down a deserted road late one night and see a light in the distance. It is moderately bright but you really have no way of knowing whether it is the bright light from a motorcycle at a distance of 1000 feet or a flashlight at a distance of 50 feet. Without knowing the type of light source, the apparent magnitude cannot tell you the distance. However, you then notice the light swing back and forth in a way that tells you it must be a flashlight. Once you have figured out the type of light source, (in this case a flashlight) you know its luminosity (absolute magnitude) based on past experience. This allows you to conclude that the distance to the source is 50 feet. The key is: if you know how bright an object *looks*, and how bright it *really is*, you can determine its **distance**.

As an example, carefully consider the case of the star Sirius. Sirius has an apparent magnitude of − 1.5, which tells you how bright it looks. It has an absolute magnitude of +1.4, which tells you how bright it really is. +1.4 is how bright it would look from a distance of 10 parsecs. Because the absolute magnitude is more than the apparent magnitude, we can conclude that Sirius would be dimmer if it were moved to a distance of 10 parsecs than it is in its true location (where it really is relative to Earth). If it would get dimmer in moving to 10 parsecs, this implies that Sirius is actually located at less than 10 parsecs from Earth. This is very powerful logic! The next step of turning this procedure into one that can give us the exact value for the distance just involves some mathematics beyond the scope of this text. The critical thing to remember is that if we could have determined Sirius' absolute magnitude and apparent magnitude without first knowing its distance, we could have found its distance without the need for trigonometric parallax.

Chapter 4
The Message of Starlight

Stars are located so very, very, far away that you or I or our great-great-grandchildren will probably never visit them. Even the light from our closest star, the Sun, requires over eight minutes (traveling at 186,000 miles per second) to complete the journey to Earth. Light from the next nearest star, Proxima Centauri, requires an astonishing 4 years to complete the journey to Earth. Yet astronomers have been able to learn an enormous amount about these distant suns by studying the starlight they emit. The primary job of the astronomer is to untangle and decode the faint beams of starlight that we receive on Earth because hidden in starlight is information about the chemical composition, temperature, size, distance, motion, and life expectancy of stars.

4.1　The EM Spectrum

When sunlight passes through a prism, a rainbow of colors appears. Sunlight (often called white light) is composed of all colors. The prism effectively separates light into its component colors. The light of each of these colors is described by a specific wavelength—the distance between adjacent crests of the waves comprising the light. Generations of astronomy students have learned the colors of the rainbow from their friend, ROY G. BIV, which stands for the first letters of the colors: Red, Orange, Yellow, Green, Blue, Indigo, and Violet. This sequence is arranged from longer to shorter wavelengths.

Box 4-1: The Nature of Light

Scientists have always struggled with how to adequately define the ethereal nature of light. Some scientists describe light in terms of small energy packets, called photons. Other scientists find it more convenient to describe light in terms of waves, focusing on characteristic wavelengths.

Astronomers tend to describe light both as particles or waves depending on the circumstances of the discussion. In terms of waves, light can be described by its wavelength or by its frequency. Light generated from a high energy or high temperature source will have short wavelengths and high frequencies. Alternatively, light generated from a low energy or low temperature source will have long wavelengths and low frequencies. For example, when wood burns in a raging campfire, the hottest part of the fire looks blue. After the fire burns out, the remaining embers, which are still very hot, but not as hot as the fire at full pace, appear red. Light waves that are mostly blue have shorter wavelengths than the longer red light waves—about 4000 angstroms for blue compared to some 7000 angstroms long for red (an angstrom is about the size of a hydrogen atom).

However, regardless of the description, light always has one characteristic that is never debated—its speed. The speed of light is nearly 186,000 miles per second (300,000 km per second). That is fast enough to make it to the Moon in less than 2 seconds! But, even at that enormous speed, it takes sunlight more than 8 minutes to reach Earth at a distance of 93 million miles. Even harder to comprehend, it takes more than 4 years for light from the star nearest the Sun, Proxima Centauri, to reach Earth. This distance that light travels over time is so well understood that astronomers often describe distances in terms of how far light travels in one year—the light-year.

Our eyes are only sensitive to a very small portion of the light being emitted by the Sun, so a more in-depth study of sunlight reveals that there is actually an entire continuum of wavelengths being emitted by the Sun. For example, the mercury in a thermometer will rise even if it is placed beyond the red edge of the rainbow. Although invisible to humans, this longer wavelength light energy (called **infrared** waves), can be sensed by a thermometer. Alternatively, just beyond the blue side of the spectrum, special instruments can detect light at shorter wavelengths even though our eyes cannot see it. This invisible short-wavelength light energy, which is just beyond the blue edge of the rainbow, is called **ultraviolet**. In fact, there is a continuum of light energy being emitted from our Sun at wavelengths both shorter and longer than the visible rainbow. This complete spectrum of light wavelengths is called the electromagnetic (EM) spectrum. It includes the longest wavelength radio waves up to the shortest wavelength gamma rays and everything in between. The portion of the EM spectrum to which our eyes are sensitive, called **visible** light, is only a very small fraction of the entire EM spectrum.

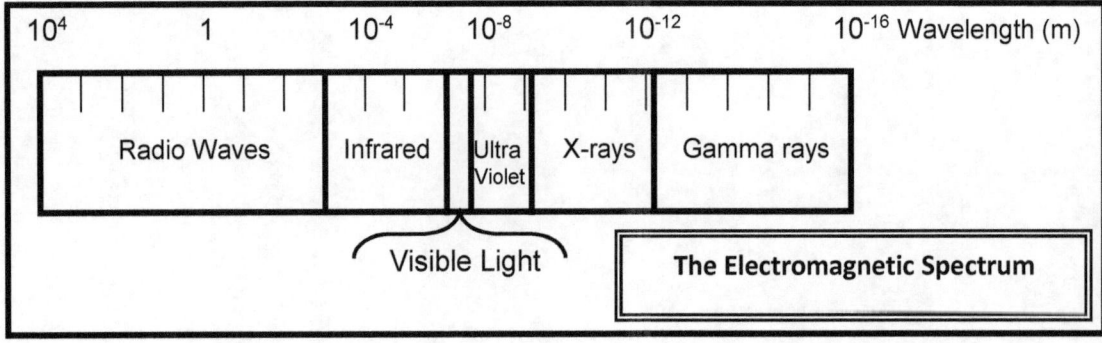

4.2 Decoding Light

Starlight arriving on Earth is a mixture of all wavelengths. Separating the light into each part of the EM spectrum is the first step to unraveling the message hidden in the mix. The astronomer's first job is to separate the light into its constituent wavelenghts so that each of the components can be examined. This is done by first passing a narrow beam of light through a prism (or diffraction grating), to produce a rainbow.

> **Box 4-2: Observing the Electromagnetic Spectrum**
>
> Our Sun and most other stars emit at least some energy at all wavelengths of the electromagnetic spectrum. Astronomers employ a variety of special telescopes designed to look at particular ranges of the electromagnetic spectrum. For example, the enormous radio telescopes at Arecibo, Puerto Rico and Greenbank, West Virginia look at the longest wavelengths of the EM spectrum whereas the giant telescopes on the mountain of Mauna Kea on Hawaii look at the infrared and visible wavelengths. However, because of the protective nature of Earth's atmosphere, the shorter ultraviolet, X-ray, and gamma ray wavelengths that must be viewed by satellites orbiting high above the Earth's surface. Representatives of these satellites include the International Ultraviolet Explorer (IUE), Chandra X-ray telescope and the Compton Gamma Ray Observatory (GRO). To accurately decode the hidden message of starlight all wavelengths of the EM spectrum must be observed in concert.

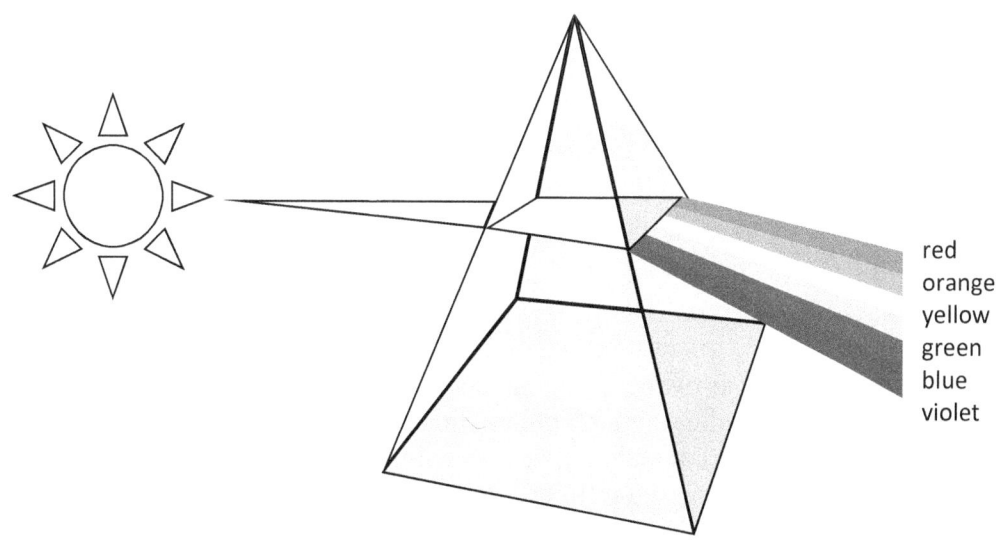

4.3 Measuring the Brightness at Each Wavelength

The next task is to measure the brightness of the starlight at each individual wavelength (color). The information about the brightness of the star at each wavelength is called the star's spectrum. Below is a table giving the brightness of the Sun at wavelengths ranging from the infrared to the ultraviolet. Another way to represent this data is to draw a graph of brightness versus wavelength. This means that brightness is plotted on the vertical (*y*) axis and wavelength on the horizontal (*x*) axis as shown below.

Color	Wavelength (Angstroms)	Brightness (arbitrary units)
ultraviolet	2500	18
violet	4200	74
indigo	4400	76
blue	4600	81
green	5300	85
yellow	5800	83
orange	6100	78
red	6600	70
infrared	9000	44

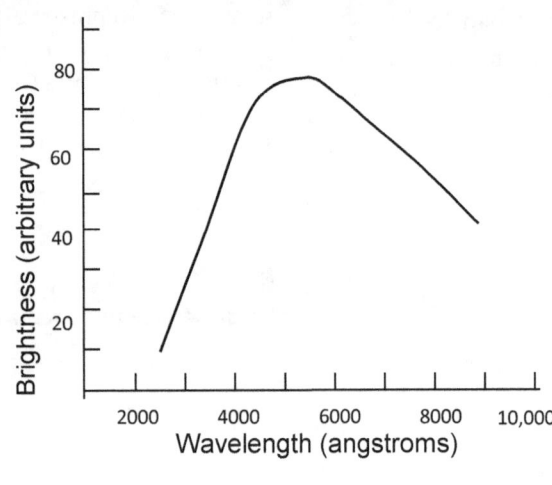

A spectrum like the one shown is officially called a **blackbody spectrum**. Hot, opaque gasses, liquids, and solids all emit blackbody spectra. The temperature of the object determines the wavelength at which the spectrum has its greatest brightness, called the peak of the blackbody spectrum. Relatively cool objects, like our own bodies, have peaks in their blackbody spectra at long wavelengths called infrared. Our bodies are too cool to emit much light in the visible range. Although our eyes cannot detect this infrared radiation, certain animals (like the pit viper snake) as well as sophisticated instruments can "see" in the infrared. Hot objects, like a stove element, emit substantial visible light and glow red. As objects become hotter and hotter, the peak in the blackbody spectrum moves to shorter and shorter wavelengths. For example, when burning a match, the hottest part of the flame looks blue. After the fire burns out, the remaining, cooler match-top appears red. Note that this is the exact opposite of what the temperature selector uses in your car because people tend to think of red as being hot and blue as being cold because those are the colors that human bodies become when exposed to such temperature extremes.

The connection between the exact location of the peak of the blackbody spectrum and the temperature of the object producing it is well understood. Astronomers can determine the temperature of a distant star by measuring the peak in the blackbody spectrum. In other words, astronomers are able to measure the temperature of a star by looking at which color is the most intense.

Stellar temperatures range enormously. The hottest stars have temperatures above 35,000 Kelvins and appear somewhat blue compared to the majority of stars (35,000 Kelvins is about $60,000^0$ Fahrenheit, which, for our purposes, is just a really big number). Stars like our Sun have surface temperatures around 6000 K. The coolest stars have temperatures of only a few thousand degrees—but still hot enough to vaporize lead! These cool stars, like Betelgeuse, really do appear red compared to most other stars.

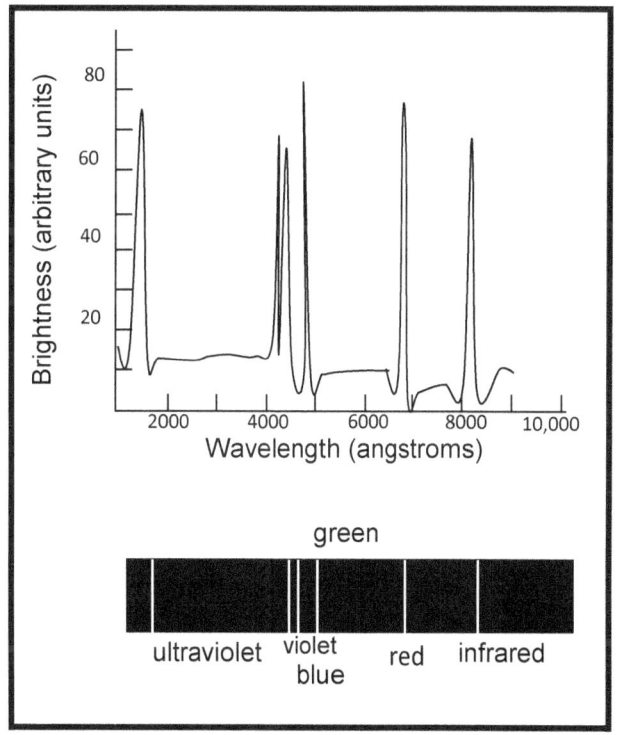

The smooth blackbody spectrum emitted by hot objects is not, however, the only type of spectrum we find in nature. If you examine the light from a hot and diffuse gas, like that in a neon sign, you will see a very different spectrum. Instead of a continuous rainbow, you will see only a series of narrow colored lines. It is almost as if someone had taken a rainbow and placed a mask over it to only let through only a small selection of the colors. This is called a **bright line spectrum** or **an emission spectrum**. An example of a bright line spectrum and the corresponding brightness versus wavelength graph is shown at right.

There are two important points to note. The first is that the lines have the same color as the corresponding positions in a continuous spectrum. The second is that the dark areas correspond to low brightness on the graph whereas the lines correspond to peaks in the graph (high brightness).

In the middle of the 18th century, two German scientists, chemist Robert Bunsen (famous for his invention of a chemistry tool known as the Bunsen Burner) and physicist Gustav Kirchhoff (noted for demonstrating that electrical impulses move at the speed of light) used the spectroscope to analyze the light emitted when various chemicals were burned. They discovered that each chemical produced its own unique bright line spectrum. The discovery that each chemical element produced its own spectral signature was a momentous breakthrough—it allowed astronomers to determine the chemical make-up (**composition**) of far away stars. Armed with this new technique, astronomers began probing the composition of all bright stars. Our own Sun had a few surprises of its own. In addition to finding hydrogen, sodium, iron, calcium, magnesium, nickel, and chromium, astronomers found a "new and undiscovered" chemical on the Sun that was not yet known to exist on Earth. They named it helium, from the Greek word for Sun, helios. Today, we know that all stars are primarily made of hydrogen and helium along with traces of the other elements previously noted.

When decoding the light from a star, we discover that there is a third type of spectrum whose origin is related to both the continuous and bright line spectrum. In 1816, German born Joseph von Fraunhofer created the first instrument to observe the Sun's spectrum. To make his spectroscope, Fraunhofer mounted a prism on a small telescope and looked at a narrow beam of sunlight coming through a narrow slit in a window shutter. He later wrote that the rainbow of colors appeared as expected; however, he saw something else. He saw an almost countless number of black, vertical lines interrupting the continuous rainbow. Today these famous lines are called **Fraunhofer solar absorption lines**. Fascinated by these unexpected lines, Fraunhofer mapped some 600 of them and gave the prominent ones alphabetical labels. Fraunhofer then turned his spectroscope to the bright stars of the night sky. He discovered that although stars have black lines in the spectrum too, the black lines' positions varied from star to star.

A black-and-white photograph of a continuous spectrum is completely white because light energy at all wavelengths exposes the film. An **absorption spectrum**, in contrast, shows a white background with a series of black vertical lines corresponding to wavelengths with no light energy—these are the absorption lines. A graph of the brightness versus wavelength of an absorption spectrum shows the familiar blackbody curve interrupted by a series of downward spikes associated with the absorption lines. The important step in beginning to understand the origin of absorption spectra was the recognition that the wavelengths of the absorption lines correspond exactly to the wavelengths of the bright lines produced by hot gasses. Today, we understand that an absorption spectrum results from the continuous spectrum created by the star having particular wavelengths removed as the continuous spectrum of light passes through the thin outer atmosphere of the star.

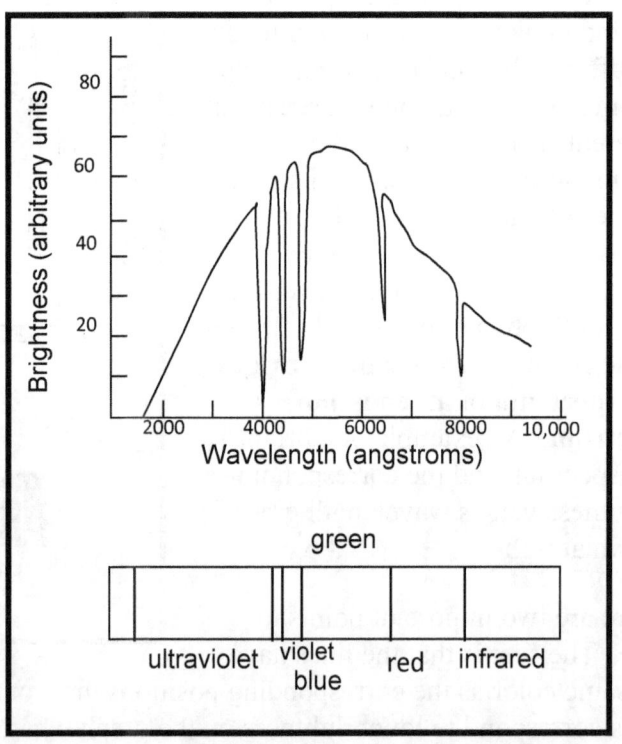

The wavelengths that are removed by the gasses in the outer atmosphere are exactly those that the same gasses would produce if heated and observed directly. By carefully noting which wavelengths are missing, astronomers can learn a great deal about the composition of the star's atmosphere. The detailed spectrum of a star is, as you can imagine, very complicated because a star contains many different chemicals simultaneously contributing to the spectrum. Today computers are used to untangle the light being emitted by each chemical.

4.4 The Message

The ability to separate light into its constituent wavelengths provides astronomers with a powerful tool for learning about stars. First, the peak wavelength in the blackbody spectrum tells astronomers the temperature of a star. Because the peak in the blackbody spectrum also determines the color of the star, astronomers often use the terms color and temperature interchangeably. A blue star is hot, a yellow star is medium temperature, and a red star is cool. Second, hidden in the complex line patterns of the absorption spectra is information about a star's composition. The otherwise smooth blackbody spectrum from every star is interrupted by a series of black lines. The positions of these lines are related to the detailed chemistry of the outer layers of the star and therefore reveal a great deal about the star's composition.

In the next unit, we will discover that there is actually an intimate connection between a star's temperature and its characteristic absorption spectrum. However, this connection is subtle and was not properly understood until the 1920s. Prior to that, astronomers assumed that any temperature star could, in principle, have any absorption spectrum. At the time there was no reason to believe in a connection between temperature and composition.

Chapter 5
The HR Diagram

In the last unit, we learned that a prism can be used to separate starlight into its constituent wavelengths yielding a surprising wealth of information about the star. The color of the star is determined by the peak wavelength in the blackbody spectrum, which is a direct measure of its temperature. However, in passing from the "surface" of a star to the surrounding space, starlight must pass through gasses that absorb specific wavelengths creating gaps in the spectrum. These gaps are known as absorption lines. Both the specific position and width of each gap provide information about the gas through which the starlight has traveled. By the late nineteenth century the absorption line spectra from a large number of stars had been photographed, revealing a wide variety of patterns.

5.1 Classifying the Stars

It was natural for astronomers to assume that the two aspects of a star's spectrum—the color and the absorption line pattern—were independent of one another as each was influenced by a different property of the star. The color was affected by the temperature whereas the absorption lines were affected by the details of the chemical abundances in the outer layers of the star. It is important to realize that the prevailing wisdom at the time was that the proportions of elements in stars were similar to those of the Earth. Differing absorption line patterns were then thought only to reveal slight variations in a chemical composition consisting mostly of iron, magnesium, silicon, and oxygen. According to this interpretation, a star with a particular absorption spectrum could exist at any temperature and therefore there should be no correlation between color and absorption line spectrum. One of the first breakthroughs in modern astrophysics was the recognition that there is indeed a correlation between a star's color and its absorption line spectrum. This was revealed primarily through the work of a number of women scientists working at Harvard University around the beginning of the twentieth century.

When scientists in any discipline are confronted with an overwhelming quantity of data, one of the first approaches to making sense of the data is to classify it. The hope is that the process of classifying will reveal some underlying pattern in the data that will eventually lead to fundamental insight. There are no specific rules for the classification process. Scientists can devote years, or even entire careers, pursuing classification schemes that ultimately prove unproductive. In the mid to late 1800s, astronomers were faced with a growing collection of stellar absorption line spectra, or simply spectra, for which they lacked any consistent theoretical explanation. In an attempt to create some order out of the apparent chaos, they began to look for similarities between the spectra from different stars in the hope that this would allow them to sort stars into groups with similar characteristics. This classification of stars would come to be known as **spectral class.**

Improvements in photography allowed astronomers to photograph the spectra of tens of thousands of stars. These were black-and-white photographs in which the spectra looked very much like the bar codes used by supermarket scanners. However, the task of cataloging and classifying ten thousand stellar spectra was tedious, often boring, and incredibly time-consuming work. So astronomers assigned this presumably low-level task to the individuals they felt were most appropriate for the task—women. Although completely inappropriate in today's society this in fact represented a unique opportunity for women to become involved in the scientific enterprise—an opportunity that they exploited to its fullest. By the late 1800s, Harvard astronomer Williamina Fleming led the development of a stellar spectra catalog describing 10,351 stars. She classified stellar spectra into 16 groups, labeled alphabetically A through Q (skipping J because it was too easily confused with I). Astronomers generally believed that stars evolved from hot to cool (blue to red) and the Harvard classification scheme reflected this. Type A stars, like Sirius and Vega, were believed to be the hottest and youngest. It was thought that these would eventually evolve into type B stars like Rigel then C, D, E, and so on. Astronomers did not believe that temperature was determining spectral class but rather that both color and chemical composition changed over the lifetime of a star.

A follow-up project led by Antonia Maury revised Fleming's scheme of 16 groups to include subgroups that distinguished subtle degrees of sharpness or fuzziness of the spectral lines. She did not know it at the time, but today we understand these characteristics to be an indication of the size of certain stars or the rapid rotation of others. More importantly, she realized that O class stars were in fact the hottest, followed by B then A.

Annie Jump Canon took over were Maury left off and made several more modifications to the classification scheme. Most importantly, she extended Maury's work on the temperature sequence to complete the accurate ordering of stars by color, from blue to red, in the sequence: O, B, A, F, G, K, M. Some astronomy students happily remember this sequence using the mnemonic: *Our Best Astronomers Feel Good Knowing More*. Others remember the sequence from blue to red as: *Oh Boy, An F Grade Kills Me*. However, by far the longest tradition has been to learn the sequence as: *Oh, Be A Fine Guy/Gal, Kiss Me*!!! Cannon was a phenomenal spectral classifier—by the time of her death in 1941, she cataloged an astonishing 395,000 spectra.

5.2 Stellar Temperatures

The great surprise that came out of Cannon's work was that there was an unmistakable correlation between a star's spectral class and its temperature. In other words, all G class stars seemed to be at the same temperature. The only explanation that seemed to make any sense was that both the temperature and chemical composition of a star change with time so that all stars that have cooled to the same temperature are made of the same thing. This correlation between temperature and spectral class was interpreted as resulting from a common cause—the evolutionary sequence. In fact, the connection between temperature and spectral class was soon discovered to be much deeper than most astronomers had dared to imagine.

It was not until 1920 that a young Indian physicist by the name of Megh Nad Saha showed how temperature can affect stellar spectra without changing the chemical composition. Saha proposed that hydrogen gas produces a different spectrum at different temperatures. Armed with this idea, Cecilia Payne-Gaposchkin was able to explain all of the spectral classes under the bold assumption that stars were mostly made of only hydrogen and helium. Although aggressively attacked by the leading astrophysicists of the time, Payne-Gaposchkin's theory eventually won out. Today we understand that all stars are made of essentially the same stuff. The differences in spectra result from changes in the way that this "stuff" behaves at varying temperatures.

What is really important about all of this is that it means we have two ways of determining a star's temperature. We can determine its temperature either from the blackbody radiation formula (by looking for the position of the broad peak in its spectrum) or we can determine its spectral class from which the temperature can be estimated. Although the first method is more direct, it is much easier to determine a star's spectral class than its peak wavelength. The use of spectral class to determine temperature therefore allowed astronomers to estimate the temperatures of thousands of stars too dim to be measured directly. Today, astronomers talk about temperature, color, and spectral class interchangeably because knowledge of any one of these properties is generally enough to tell us the other two.

5.3 Stellar Sizes

The ability to determine the chemical composition and temperature of distant stars is quite impressive in itself, but when coupled with the ability to use parallax to measure the distance to stars, another piece of information can be decoded from starlight—the size of a star. Two Austrian physicists, Josef Stefan and Ludwig Boltzmann demonstrated that a star's luminosity is directly related to its temperature. This important relationship between luminosity and temperature is known today as the Stefan-Boltzmann law. Although the mathematical details are unimportant for this discussion, there are two vital aspects that must be understood:

1. The total amount of light energy that a star emits—called the luminosity and measured by the absolute magnitude—increases with temperature. In fact, it increases as the **fourth power** of the temperature so that a star the same size as our Sun but twice the temperature would be 16 (or 2^4) times more luminous.

2. The amount of energy that a star emits per acre of surface area depends only on the star's temperature. Therefore, the total luminosity of a star increases with the number of surface acres. A star at the same temperature as our Sun but four times the surface area is four times as luminous. (Surface area of a sphere is given by $4\pi r^2$.)

To estimate the size of a star, we first determine its temperature from either its color or spectral class. This tells us how much energy each acre of the surface is emitting. Knowing the total luminosity determines the total number of acres of surface area, which is a direct measure of the star's size. An example is shown in Box 5-1.

> **Box 5-1: Determining the Relative Size of a Star**
>
> **Q:** *A star is observed to have 64 times the luminosity of the Sun ($64L_{sun}$). A measurement of its spectral class reveals that it has a surface temperature that is twice that of our Sun. What is the radius of the star compared with the radius of the Sun?*
>
> From step one we know that the amount of energy per unit of surface area will increase 16 times for a doubling of the temperature. This means that this star must emit 16 times as much light for each acre of surface than the Sun does. But, if it is emitting 64 times more energy in total, it must have 4 times the area of the Sun (4x16=64). If something has 4 times the surface area, it must be twice as far across. Therefore, this star must have a radius that is twice the radius of our own Sun. Although the mathematics may become a little more complicated, this same method can be used to estimate the size of any star.

5.4 The HR Diagram

By decoding starlight, astronomers can deduce a star's luminosity, temperature, and size without ever leaving our Earthly confines. However, which, if any, of these stellar characteristics are related? The answer to this question was first proposed in the early part of this century when two researchers, amateur astronomer Enjar Hertzsprung and Harvard professor Henry Norris Russell, independently developed what has become known as the **HR diagram** (facing page). They made a plot showing stars' temperatures on the horizontal axis and corresponding luminosities on the vertical axis, which revealed some interesting relationships. Nature appears to prefer certain combinations of luminosity and temperature to other combinations.

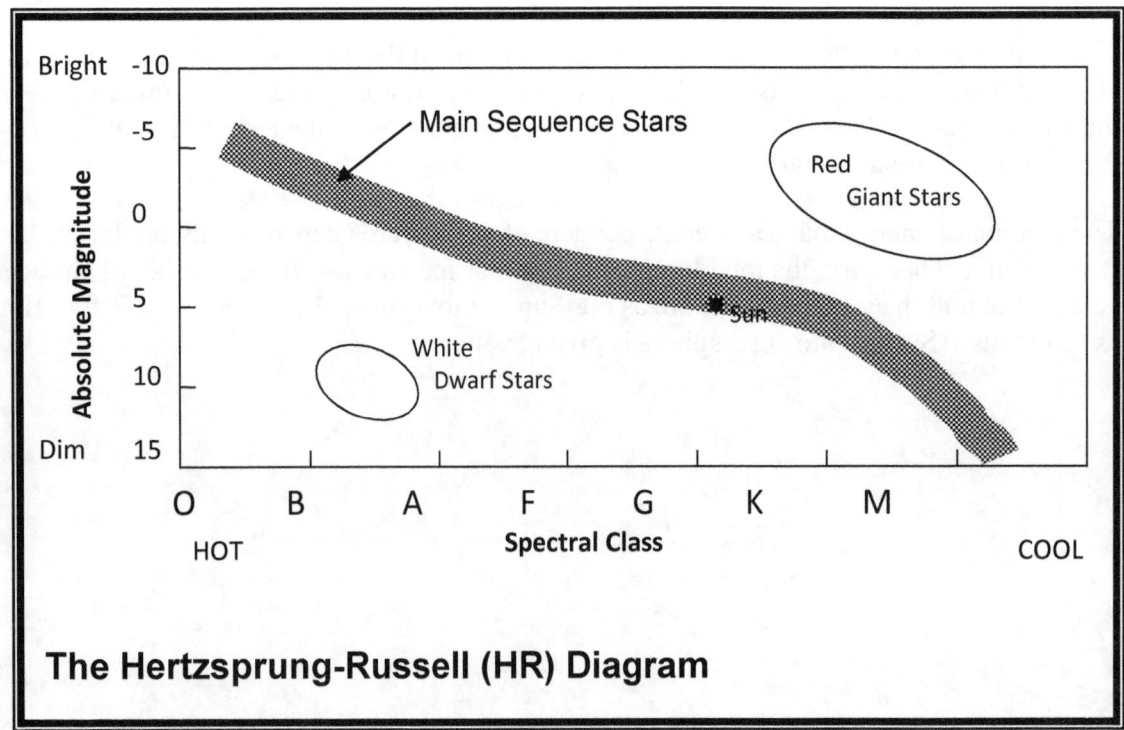

The Hertzsprung-Russell (HR) Diagram

Both Hertzsprung and Russell were surprised to find that almost every star was located on a band going from the upper left to the lower right part of the HR diagram. There were, however, a few rogue stars that did not fit this pattern—they mostly appeared in the upper right corner with just a few showing up in the lower left corner. We will consider each of these groups of stars on the HR diagram one at a time.

The Main Sequence

The most prominent grouping on the HR diagram is the line that goes from the top left (large, bright stars) to the bottom right (small, dim stars). This grouping is called the main sequence and includes (as you can see in the figure) our own Sun. Most stars in the galaxy are main sequence stars.

The Red Giants

The upper right corner of the diagram contains a collection of stars that, though cool, are tremendously bright. The amount of light emitted per unit of surface area is small for a cool star and therefore the only way to achieve such enormous luminosity is for the star to be very large. Consider the star Betelgeuse in the constellation Orion. This red star has a surface temperature of about 3500 K as compared to about 5600 K for our Sun. The Stefan-Boltzmann law suggests that Betelgeuse emits only about 10% as much light per acre of surface as the Sun. But, Betelgeuse has a luminosity of about 100,000 L_{sun}, which means that it must have about one million times the area! This means that Betelgeuse must have a radius approximately 1000 times that of the Sun—a true giant.

The White Dwarfs

A small group of stars are found in the bottom left of the HR diagram and are called white dwarfs. These are extremely hot stars but have a low luminosity. Therefore, they cannot be very large. For instance, Sirius B is only about 3% as luminous as the Sun ($0.03 L_{sun}$) but is the same temperature as Spica, a main sequence star that is about 10,000 L_{sun}. Spica is therefore about 300,000 times more luminous than Sirius B. As these two stars have the same temperature, Sirius B must be considerably smaller—about 1/600 of the radius, which makes Sirius B considerably smaller than the Sun. We usually think of white dwarfs as being about the size of the Earth.

5.5 Spectroscopic Parallax: A New Method to Determine Distance

We have already discussed how the method of stellar parallax is only applicable to stars within a distance of 250 parsecs—stars any farther away do not exhibit enough parallax to be measured. Fortunately, the discovery of the relationship between luminosity and temperature for main sequence stars provides astronomers with the next rung in the distance ladder. If a main sequence star's spectral class is known, then it is a reasonable assumption that it will have the same luminosity as all of the other stars in that spectral class. Then, using the HR diagram, the luminosity can be determined. For example, our Sun is a G2 star. As shown on the HR diagram, an imaginary line drawn vertically upward from G2 to the main sequence corresponds to a value of 4.8 for absolute magnitude. As described in the Standard Candles section, the difference between a star's apparent and absolute magnitudes can be used to determine distance. Using the HR diagram to determine distance is called spectroscopic parallax. It is a poor name because it has nothing to do with parallax and measuring angles. This method requires four principal steps:

1. Measure the star's apparent magnitude.

2. Acquire the star's spectrum and identify its spectral class.

3. Use the HR diagram to find the star's corresponding luminosity.

4. Compare the apparent and absolute magnitude to determine the star's distance.

This method works quite well if the star is a main sequence star. However, the calculated distance will be woefully inaccurate if the star is not on the main sequence. However, back in 1897, Antonia Maury noticed some stars have sharp and narrow gaps in the absorption spectra—today we know these are giant stars. Similarly, there are also clues in the spectrum that indicate dwarf stars. Using these subtle clues, astronomers can be sure to only apply spectroscopic parallax to finding the distance to main sequence stars.

> **Box 5-2: Self Check**
>
> If an astronomer uses the spectroscopic parallax method to determine the distance to an M class star and mistakenly identifies it as a main sequence star when it is really a red giant, will her estimate for the distance be too small or too large?

As a simple example, imagine that you measure the spectrum of the main sequence star Hadar and find it to be a type B1. From your library research, you know that the main sequence star Spica, located at a distance of 70 pc, is also a B1 and therefore the same temperature as Hadar. Standing outside, you measure Spica to appear 4 times brighter than Hadar. Therefore, for Spica to appear four times as bright it must be at half the distance, which puts Hadar at a distance of 140 pc. Measurements of only the spectral class and apparent magnitude of Hader allowed its distance to be determined. Of course this conclusion rests heavily on the assumption that the luminosities are identical (which will not always be exactly the case), because the main sequence is not a highly restricted curve.

Chapter 5
Stellar Life Cycles

6.1 Star Dust and Protostars

It might surprise you to know that the open space between the stars is not really empty. Outer space is filled with just a few specks of dust and gas—matter that astronomers call the interstellar medium (ISM). The ISM is composed mostly of hydrogen and helium like stars but has a much smaller density. In fact, a cubic inch of the ISM only contains a few isolated atoms. Occasionally, these pieces of dust and gas gather together in large clouds called interstellar clouds or molecular clouds where the number of atoms per cubic inch is slightly higher than the ISM. Here it can be on the order of 1000 atoms per cubic inch.

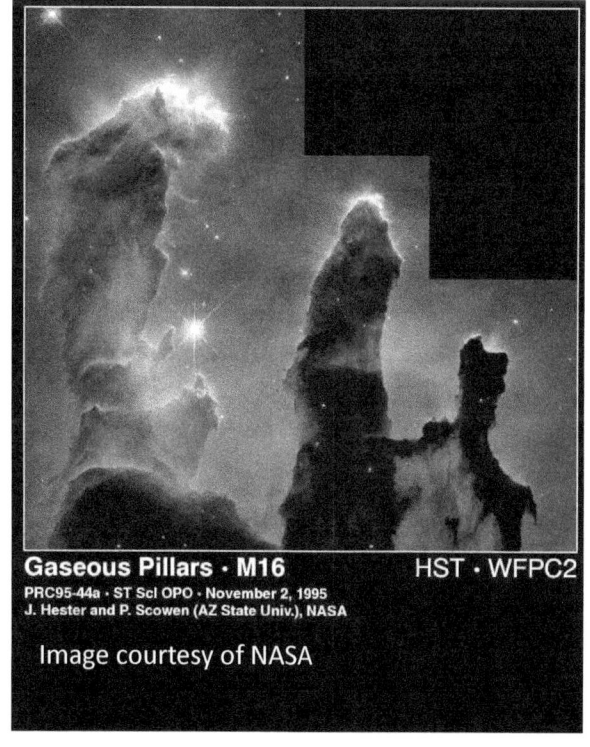

These molecular clouds are quite easy to observe. One place is to look is the faint band of light stretching overhead called the Milky Way. As viewed from Earth, the Milky Way is a swath of millions of stars—many more than can be counted easily. Close inspection will reveal what appear to be dark regions of few stars. These dark areas, called dust lanes, are actually the result of nearby large molecular clouds obscuring the background stars. Another place to look is in the photographs of nebulae taken by large astronomical observatories. A photograph of the great Orion Nebula shows dust and gas being illuminated by the bright stars nearby. A close-up view of the Pleiades star cluster, in Taurus, shows dust and gas left over from the formation of some 300 hot, young stars. And, of course, NASA's Hubble Space Telescope has made some great images of molecular clouds. One of the best pictures is that of gaseous pillars in the Eagle Nebula in M16.

Box 6-1: Rotation of Protostars, Stars and Planets

When molecular clouds collapse to form protostars, the inward falling of material is not evenly distributed on all sides of the central core. This uneven collapse induces a small amount of spin that eventually has an enormous impact on the star. Because of a law of physics called conservation of angular momentum, the collapsing cloud will begin to spin faster and faster as it contracts. This is the same principal that is evident when an Olympic ice skater, who is initially spinning slowly, pulls his arms inward and begins to spin faster. The initial clockwise or counter clockwise spin of a protostar is retained by the formed main sequence star and dominates the direction planets will orbit the central star. For example, our Sun spins counter clockwise, as viewed from Polaris, about every 28 days. All of the planets and asteroids in our solar system orbit the Sun in the same counterclockwise direction and most planets even rotate counterclockwise. All of these motions are explained by the conservation of angular momentum law from physics.

In the picture of M16, the pillar is about 1 light year long but some molecular clouds are as large as 100 light years across! Some of these enormous molecular clouds collapse inward. Astronomers do not know exactly why this occurs—maybe the cloud is super-massive and collapses under its own weight or maybe the shock wave from a nearby exploding star causes it to collapse. Regardless, a collapsing cloud will heat up in the same way that compressed air in a bicycle pump becomes hot. This pre-star collection of material is known as a **protostar**. Eventually the central density and pressure are high enough that nuclear reactions begin. Small atoms combine to form more massive ones and energy is released. This reaction is called **nuclear fusion**. When the molecular cloud has collapsed into a protostar and the central core is hot and dense enough to sustain nuclear fusion, the protostar becomes a stable **main sequence star**. Early scientists thought that the stars were powered by either coal burning or gravitational contraction. Today, however, we understand that the 20 million degree temperatures at the center of stars is great enough that hydrogen atoms combine to form helium and, in the processes, release energy which eventually makes its way to the star's surface as EM radiation (light).

6.2 Main Sequence Stars

Stars spend most of their lives as main sequence stars, quietly converting hydrogen into helium. Two simultaneous processes dominate all main sequence stars: the force of gravity pulling the star inward and the outward pressure of energy escaping the core. The perfect balance of inward pushing gravity and outward pushing radiation pressure is called **hydrostatic equilibrium**. The main sequence phase continues until the star's core begins to run out of fuel.

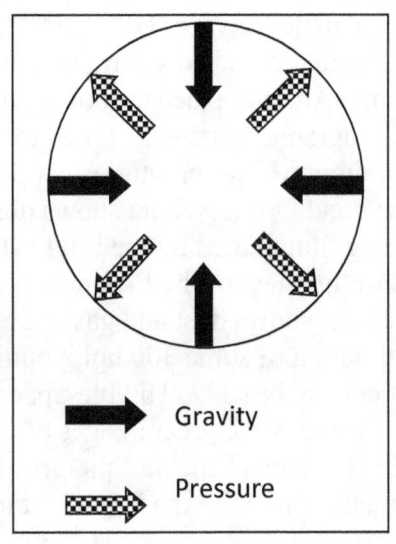

How long it takes for a star to use all of its available fuel depends only on two variables: (1) how much fuel is available for nuclear fusion and (2) how fast the star burns its fuel. As you might predict, the brighter a star shines, the faster it has to convert hydrogen into helium. This conclusion is actually a direct result of hydrostatic equilibrium. An incredibly massive star has lots of gravity trying to collapse it; therefore, the interior of the star has to burn fuel at an enormous rate to create a balancing outward radiation pressure and, subsequently, maintain hydrostatic equilibrium. For example, a super-massive class O star with a mass of $60M_{sun}$ will exhaust all its fuel in less than a million years. A star of 10 solar masses can be expected to live for 10 million years while our Sun formed with sufficient fuel to continuing burning for about 10 billion years. On the other hand, small class M stars, which do not burn fuel very fast at all, can live upwards of 50 billion years. In fact, if current estimates of the Universe's age of 12 billion years are correct, then the Universe is not old enough for any M class stars to have run out of fuel and left the main sequence! Generations of astronomy students have remembered the phrase, "**burn bright— die young**" as the mantra for recalling stellar lifetimes.

Our Sun is converting hydrogen to helium at a rate of about 6×10^{11} kg of hydrogen—that's about the same amount of material that is in Mt. Everest—every second! Although that seems like a large amount of hydrogen, our Sun is more than 109 times the diameter of Earth and has at least enough hydrogen to last a total of about 10 billion years. Current estimates place the age of the Sun at about 5 billion years, thus making it a middle aged star.

When the Fuel Tank is Empty

A variety of things happen to stars when they begin to run out of fuel, the details of which depend entirely on the star's mass. In general terms, there are two tracks for the death throes of stars: (1) low mass stars—stars much like our own Sun—follow a track that leads to the formation of white dwarfs and (2) high mass stars—stars with masses more than five times the mass of our Sun—follow a track that leads to the more exotic neutron stars and black holes. Let us consider each type individually.

6.3 Life Cycle of a Low Mass Star

Eventually, the core of a star has little hydrogen left because most of the atoms have been converted to helium. At this point the nuclear fusion in the core stops. Helium requires 100 million degree temperatures to sustain nuclear fusion as opposed to the cooler 20 million degree temperatures that hydrogen fusion needs. The core becomes unstable and begins to shrink. This heats the hydrogen surrounding the core to temperatures sufficient for hydrogen fusion to occur in a thin shell around the core. This shell-source fusion causes the nearby outer atmosphere of the star to heat and expand. This expansion causes the star to grow and become brighter. The star is now a **red giant** star— bright because it is large and red because the outer layers are cool as a result of being far from the hot stellar core.

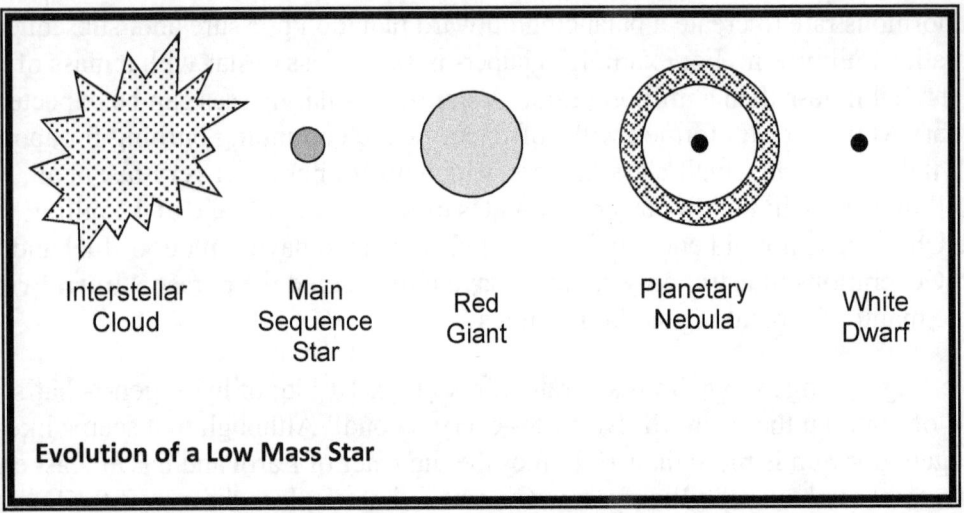

Evolution of a Low Mass Star

A red giant star is inherently unstable. Eventually the outer atmosphere detaches from the hot inner core. The result is a hot and dim star (called a **white dwarf**) surrounded by an expanding shell of gas. This thin shell of gas, called a **planetary nebula**, glows on its own because it is being bombarded by the white dwarf's ultraviolet radiation. Eventually this thin shell dissipates and leaves only the white dwarf behind to slowly (over tens of billions of years) cool until it no longer emits light. Nuclear reactions are no longer occurring in the dwarf—its only light source is from cooling. Much like a stove element shines red while it cools, the dwarf will eventually cool to the point that it is no longer visible, and what remains is just a lump of primarily helium and/or carbon ash left over from a star. This object will some day be known as a **black dwarf**. Five billion years from now, this too will be the fate of our Sun; the Earth, eventually being dragged down to the core of the red giant, will cease to exist.

Novae and Type I Supernovae

When interstellar molecular clouds are large enough, they sometimes condense in several areas forming a multiple star system. These binary star systems are composed of two stars, bound together by gravity, orbiting around a central point between them. It is currently thought that as many as one third of all stars in the sky are actually part of multiple star systems.

If one of two stars in a binary system is more massive than the other, the more massive one will burn its fuel faster, evolve faster, and become a white dwarf before its partner finishes its main sequence phase. When the second star becomes a red giant with a large and tenuous atmosphere, the gravitational attraction of the white dwarf can actually steal material from its companion red giant. In this situation the white dwarf will begin to grow.

In 1928, an Indian graduate student named Subrahmanyan Chandrasekhar set sail from India to study at Cambridge University in England. During his voyage, he considered the problem of a white dwarf attempting to support its own crushing weight without the benefit of nuclear fusion in the core to supply radiation pressure pushing outward. In this case, the only thing preventing the star from collapsing would be the forces between the individual atoms. What he discovered is that only a white dwarf with a mass less than about $1.4 M_{sun}$ can support itself against the force of gravity and maintain its size. If however, the white dwarf grows to more than $1.4 M_{sun}$ then it is no longer able to support itself and collapses under its own weight with a tremendous release of energy. Today this limiting mass of 1.4M_{sun} for a white dwarf is known as the **Chandrasekhar limit** and plays a vital role in determining the ultimate fate of growing white dwarfs scavenging material from their binary companions.

The material that the white dwarf steals from the outer layers of its binary companion is primarily hydrogen—the primary fuel for nuclear fusion. As the hydrogen piles up, the pressure at the base of the hydrogen layer grows and the hydrogen heats up. If the conditions at the base of the hydrogen layer reach the critical threshold for nuclear ignition before the total mass of the white dwarf exceeds the Chandrasekhar limit then this outer layer can ignite. The surface of the white dwarf can shine 10,000 times brighter than usual for several days as the hydrogen on the surface burns off. Depending on how fast the hydrogen piles up this can repeat every 10,000 to 100,000 years. Astronomers call this repeating event a **nova**, which is Latin for "new star." Alternatively, if the mass of the white dwarf is initially close to the Chandrasekhar limit then it is possible for the white dwarf to become too massive to support itself. The result is an explosion millions of times brighter than the original brightness of the perished white dwarf leaving virtually no remains. This catastrophic event is called a **type I supernova**.

6.4 Life Cycle of a High Mass Star

Stars that begin larger than about 5 M_{sun} follow a faster and more violent life cycle than that of low-mass stars. Like low-mass stars, high-mass stars also form from large molecular clouds by first forming protostars followed by stable main sequence stars. These hot and massive stars reside in the upper left-hand corner of the HR diagram and shine brightly. To balance the enormous pull of gravity trying to collapse the star, the core must quickly fuse hydrogen into helium to maintain hydrostatic equilibrium. Hydrogen in the core is quickly depleted and the star becomes unstable, just the same as we saw for a low-mass star. The enormous size of a high-mass star will also cause them to die a different type of death than that of low-mass stars.

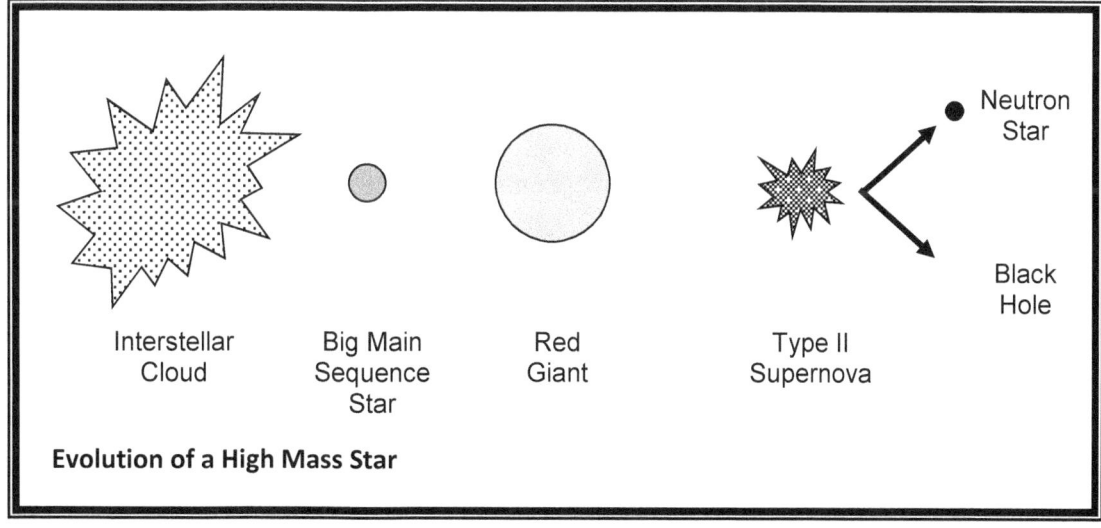

The voluminous amount of material that composes a high-mass star creates unbelievably high pressures and temperatures of more than 100 million degrees in the central core. When the star depletes its hydrogen fuel, the core is hot enough for helium-ash to fuse together to form even heavier elements—beryllium and carbon. These elements in turn fuse together at high temperatures to form oxygen, neon and magnesium. Subsequently, these elements fuse to form silicon, chromium, manganese, cobalt, nickel and finally iron atoms at 800 million degrees. For example, (in round numbers) a star of 20 M_{sun} burns hydrogen for 10 million years, helium for 1 million years, carbon for one thousand years, oxygen for one year, and silicon for a week.

This nuclear fusion process suddenly ceases with the creation of iron because to fuse iron atoms with other atoms requires an input of energy instead of releasing energy. The nuclear fusion in the hot iron core stops. At the conclusion of the nuclear fusion chain, the iron core, no longer supported by radiation pressure pushing outward, collapses in a matter of seconds. The outer layers of the star collapse downward onto the stellar core at more than 50,000 miles per second and bounce off the core in a supersonic shock wave observed as a gigantic explosion. In less than one second, the explosion emits more than 100 times the total energy the Sun will emit over its entire 10 billion-year lifetime! This dramatic end of a high-mass star is called a **type II supernova**.

Stellar Remnants: Neutron Stars and Black Holes

The expanding shell moves away from the supernova core at velocities of more than 60,000 miles per second and plows through the interstellar medium causing the ISM to heat up. The result of this collision is to create wispy and chaotic clouds called a **supernova remnant**. Although supernovae are predicted to happen at a rate of about 3 per century in the Milky Way, they generally occur in regions that we cannot see. The most recent nearby supernova is known as Supernova 1987A, which exploded in the Large Magellanic Cloud (LMG), a nearby dwarf galaxy. Although we saw this explosion in 1987, the LMG is about 165,000 ly away, so the star actually exploded about 165,000 years ago!

The remains of a type II supernova are indeed exotic. For stars with masses between 5 and 25 times larger than the mass of our Sun, the stellar core is crushed by an implosion. The collapse is so violent that the protons and electrons that composed the iron atoms are smashed together to form subatomic particles called neutrons. The resulting core is now composed of nothing but neutrons and is called a **neutron star**. Neutron stars are quite small, sometimes less than 10 miles across. Yet, the entire mass of the previous star's core is compacted therein so it is very dense. So dense that a teaspoon of neutron star material would weigh more than a ton!

Although the existence of neutron stars had been proposed in the 1930's, they were not knowingly observed until 1967—and even then it was by accident. At that time radio astronomers were looking at objects known as quasi-stellar objects, or quasars for short. A graduate student, Jocelyn Bell working at Cambridge University, was making regular observations of what she thought was a particularly uninteresting quasar. She noticed that the object emitted a series of regular pulses every 1.3373011 seconds—more accurate than the best atomic

clocks ever created. No one, including Bell, could imagine what natural phenomenon could produce such precise pulses. Astronomers jokingly began referring to this mysterious object as an LGM—a radio source from a civilization of **L**ittle **G**reen **M**en. Eventually, three more of these clock-like objects with pulsed radio sources were discovered by Bell. Certainly, there could not be four different extraterrestrial civilizations trying to communicate with Earthlings. The name **pulsar** was eventually adopted for these sources of pulsating radio emissions. In fact, when astronomers looked at the Crab nebula, they discovered a pulsar emitting radio pulses at 30 times every second.

Today, astronomers recognize that pulsars are really just rapidly spinning neutron stars. Neutron stars, or pulsars, have an enormous magnetic field. As the pulsar spins, charged particles, particularly electrons, are emitted from the north and south magnetic poles. If this beam of particles is oriented so that observers on Earth can see it, we see a flash, or pulse, of energy every time the star spins. This idea is sometimes called the "light house" model of a pulsar because of the analogy to the rotating light that warns ships of threatening rocks or signals airplanes where to land safely. To date, approximately 300 pulsars have been discovered with rotational periods as fast as 1000 rotations every second.

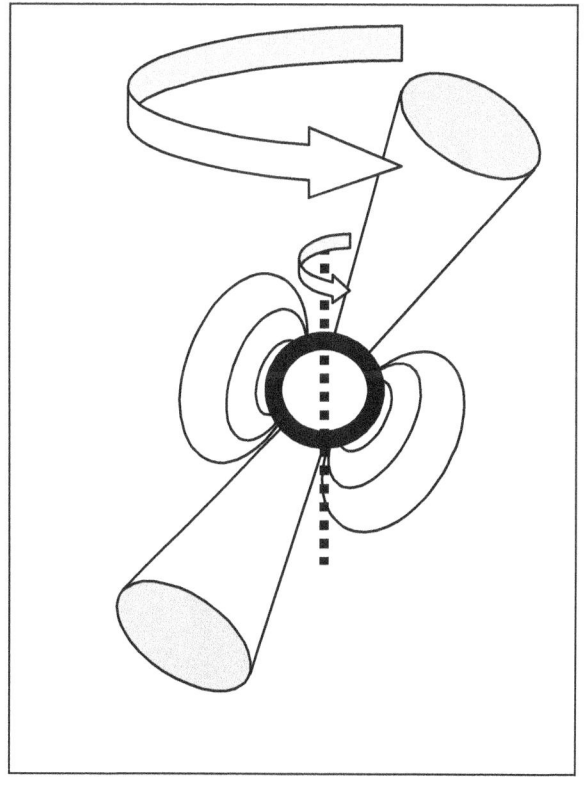

Thus far, we have described the end-state for low mass stars (about $1M_{sun}$) and high mass stars (from $5M_{sun}$ to $25M_{sun}$). However, the end-state of a super massive star, one that is more than $25M_{sun}$, is unimaginably bizarre. When a super giant star greater than $25 M_{sun}$ goes supernova, the implosion of the star onto the core is so strong that nothing can stop the resulting collapse. In fact, the core collapses forever.

The space remaining where the 25 M_{sun} star once resided is known as a **black hole**. It is an unimaginably dense and small region. Actually, a black hole is not a hole at all; rather, it is the tiny spherical remains of the stellar core. It has so much mass compressed into a small volume that its gravity is enormous—so strong that not even light can escape. Of course, an object that emits no light can be difficult to observe at best. However, if a black hole forms from one of the stars in a binary star system, then material from the remaining star can be drawn into the scavenger black hole. The material that is being piled on top of the black hole will emit energy in the form of X-rays just before disappearing from our view.

So, the trick to seeing a black hole is to look for a mysterious emission of X-rays not accompanied by a visible object. There are a number of such locations found in the sky. These mysterious X-ray sources include Cygnus X-1 in the constellation of Cygnus, A0620-00 in Monoceros, and LMC X-3 in the Large Magellanic Cloud. With the launch of NASA's Advanced X-ray Astronomical Facility (AXAF) satellite (called *Chandra* after Subrahmanyan Chandrasekhar), many more black holes have been discovered. In fact, Chandra has looked extensively at *our* galaxy, teaching us much more about the super-massive black hole named Sagittarius A* and other strange inhabitants at the center of the Milky Way.

Chapter 7
Galaxies and the Universe

Hubble Space Telescope image of the rich galaxy cluster, Abell 2218.
Credits: W. Couch (University of New South Wales), R. Ellis (Cambridge University), and NASA.

In the eighteenth century, French comet-hunter Charles Messier (1730-1817) had a problem. As he scanned the night sky with his telescope looking for undiscovered comets, he kept finding stationary fuzzy objects that could easily be confused with faint comets. He kept a record of these dim, non-stellar objects that became known as the Messier Catalog. Each of the 110 Messier objects is referred to as M1, M2, M3, and so on. Many of these Messier objects also have common names. For example, M31 is the Andromeda Galaxy and M42 is the Orion nebula. Today we understand that galaxies are far beyond our home Milky Way Galaxy and nebulae are actually inside our galaxy, but it was not until 1924 that astronomers knew this for certain.

7.1 Classifying Galaxies

Without a doubt, one of the most impressive vistas through a telescope is that of a galaxy. A **galaxy** is an organized system of stars ranging in number from hundreds of millions to upwards of thousands of billions. These vast collections of stars are also sometimes mixed with interstellar dust and gas. A quick survey of a telescope image showing a cluster of galaxies shows that galaxies come in all shapes, structures, sizes, and even colors. Astronomers typically group galaxies into four types based only on their appearance: elliptical galaxies, spiral galaxies, barred spiral galaxies, and, of course, a catchall group called irregular galaxies. Irregular galaxies make up about 10% of the known galaxies in the Universe.

In the 1920s astronomer Edwin Hubble developed a classification scheme that has come to be known as the **Hubble Tuning Fork Diagram** because of its similarity to a musician's tuning fork. In this scheme, he differentiated among the elliptical galaxies based on how eccentric, or squashed, they appear. Using a scale of 0 to 7, he designated perfectly round elliptical galaxies as E0 and highly eccentric (elongated) elliptical galaxies as E7. Additionally, he sub-classified spiral and barred spiral galaxies as (a), (b), or (c). In this scheme, (a) class galaxies had tightly wound arms and a large, bright nucleus whereas (c) class galaxies had loosely wound arms and a relatively small nucleus. Just to be complete, he classified galaxies that fell between elliptical galaxies and spiral galaxies as S0, which today are known as lenticular galaxies.

Edwin Hubble presented this simple, yet powerful, classification scheme for the purpose of advancing his theory of galaxy evolution. He proposed that all galaxies initially form as perfectly round elliptical galaxies and, over time, become elongated and develop spiral arms. If this were indeed the case then we would expect elliptical galaxies to contain mostly young stars and spiral galaxies to be composed predominantly of older stars. In fact, the situation is quite the reverse. Elliptical galaxies are generally composed of only older stars while spiral galaxies contain a mix of old and young stars. Irregular galaxies appear to contain mostly young stars. The issue of the formation and evolution of galaxies has proven to be a much more complicated and demanding problem than Hubble ever imagined.

7.2   The Milky Way Galaxy

Our Sun is one of about 100 billion stars that compose our home galaxy, the Milky Way. The Milky Way gets its name from the faint band of light that stretches overhead in the summertime night sky. When you look at this swath of stars in the sky, you are looking along the plane of the Milky Way, which is where most of the stars are located. In fact, as you look in the direction of Sagittarius, you are looking directly at the center of our galaxy.

Box 7-1: Hubble's Galaxy Classification Scheme

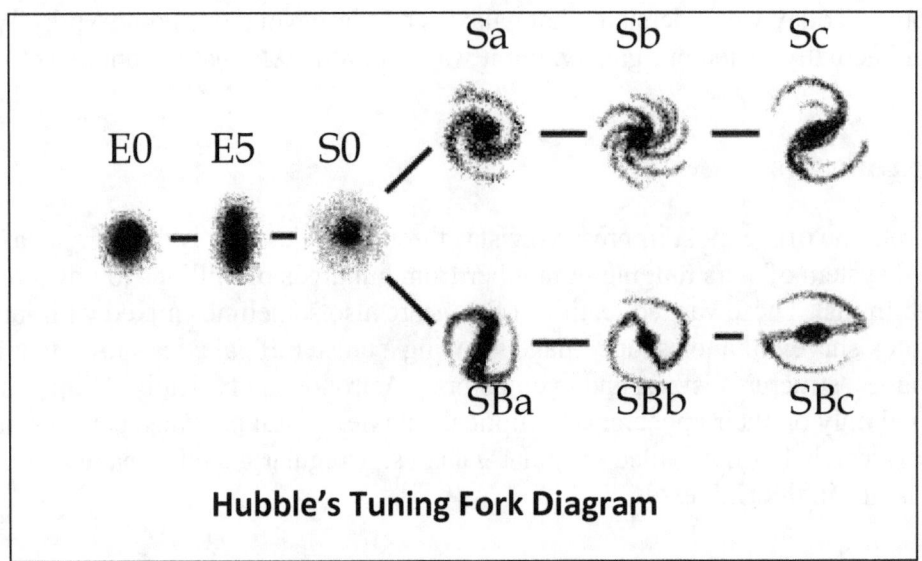

Elliptical Galaxies: An elliptical galaxy shows no spiral structure and can vary from almost round (what Hubble called E0) through to cigar shaped (called E7). This classification is based on our perspective from Earth and not on the actual shape.

Spiral Galaxies: As their name implies, spiral galaxies have outstretched, curving arms suggestive of a whirlpool or pinwheel. Hubble distinguished different sub-classes according to the tightness of the arms and the size of the nucleus. He called these Sa, Sb, and Sc. In terms of the arms, Sa is the tightest wound while Sc is the most open. In terms of the nucleus, Sa has the largest while Sc has the smallest. The galaxies that appear to have a spiral disk but no visible arms are called S0.

Barred Spirals: This galaxy type shows the same spiral structure as normal spirals, but also show a prominent bar through the nucleus. The spiral arms emerge from the end of the bar. The sub-classifications are the same as for normal spirals.

Irregulars: Certain galaxies lack either an obvious spiral structure or nuclear bulge but instead appear as a random collection of stars with no obvious order. They are distinguished from ellipticals by their lack of symmetry.

There are, of course, no photographs of the Milky Way Galaxy because there is no way to get outside the galaxy to look back in. The problem is similar to trying to figure out what the outside of your house would look like without ever going outdoors. However, if you could not go out, your first guess might be that your house does not look that different from the one you can see across the street or out the side window. In the same way, the first way to get a feel for what our galaxy looks like is to examine our neighbors. Based on a number of observations about the distribution of material within our own galaxy, astronomers have concluded that our galaxy is a medium-large barred spiral galaxy much like the Andromeda Galaxy (M31). The galactic disk extends about 20,000 parsecs out from the center—we usually write this as 20 kpc (20 kiloparsecs). The disk is about 300 parsecs thick and contains the characteristic pinwheel structure that makes ours a spiral galaxy. The galaxy features a central bulge called the galactic nucleus and globular clusters surrounding the galactic center like a halo. A globular cluster contains up to a million stars in a relatively small spherical region.

Surprisingly, the distribution of stars throughout the disk is nearly uniform—stars are not concentrated along the spiral arms. It turns out that the spiral arms are actually defined by the dust, hot O and B stars, nebulae and supernovae that reside there. In fact, it is likely that the concentration of dust in the spiral arms makes these regions particularly active in promoting star formation. Elliptical galaxies, which have no spiral arms, lack the younger O and B stars that can only exist in active star-birth regions.

Perhaps unsettlingly, our Sun is not at the center of the galaxy. Our Sun is located about 8.5 kpc (about half way) from the center on the inner edge of the Orion arm. We know this because we see that the globular clusters (which hover around the center of our galaxy) are only visible in one part of the sky near the stars of Sagittarius. If we were located in the center of the galaxy, then we would see globular clusters in every direction.

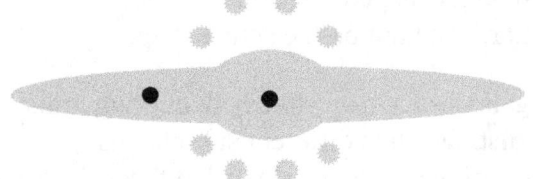

An observer in the center of a galaxy would be surrounded by globular clusters. On the other hand, an observer off to one side of the galaxy would only see globular clusters in one particular direction. It is this second observation that we make from Earth.

Our whole galaxy is spinning about the center, much like a pinwheel. At a distance of 8.5 kpc from the galactic center, it takes our Sun about 200 million years to orbit around once. This means that our 5 billion-year-old Sun has orbited the galactic center about 25 times.

7.3 Measuring the Distance to Galaxies

In the winter of 1924 Edwin Hubble made a discovery of such importance that his name is used on the first visible light space telescope. He was able to identify variable stars known as Cepheid variables in the Andromeda Galaxy more than 2 million light years away. Back in 1912, Henrietta Leavitt had already shown that the brightness fluctuation of Cepheid variables is closely related to their absolute magnitude, or luminosity. Cepheid variables that take a long time to go through their bright-dim-bright cycle have a much higher absolute magnitude than Cepheid variables that fluctuate more quickly.

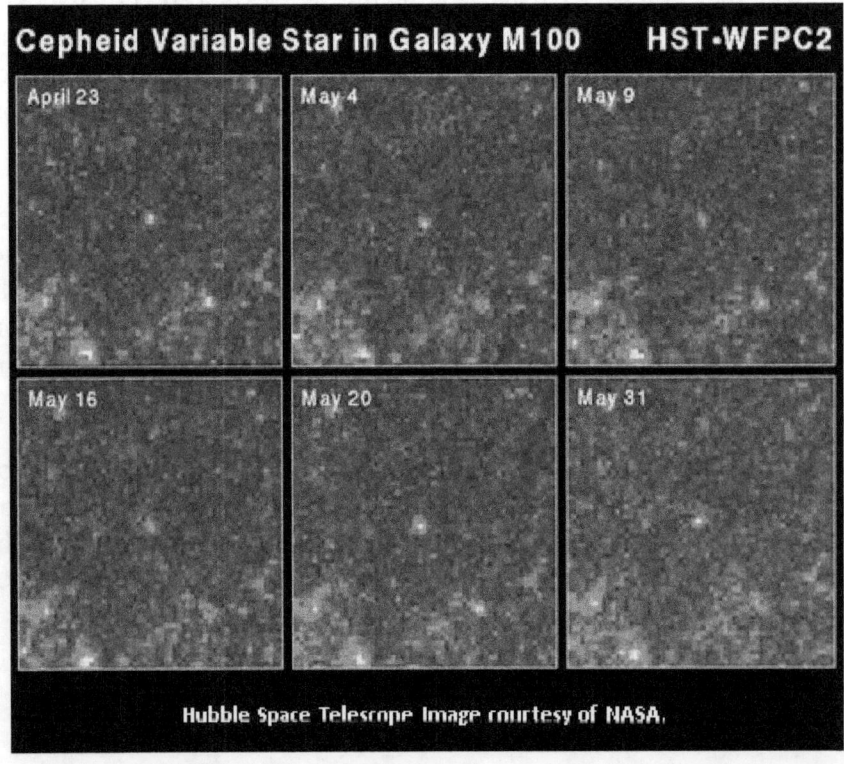

In other words, if astronomers can measure the time it takes to go through a complete brightness variation cycle, called the period, they can determine the star's absolute magnitude. This is critically important because if the absolute magnitude is known and the apparent magnitude is measured, (using the standard candles method) then the distance can be determined. By identifying Cepheid variables in halo globular clusters surrounding the Andromeda Galaxy, Edwin Hubble was able to accurately measure the distance to the Andromeda Galaxy.

Since then, astronomers have been able to map out the structure of the Universe by measuring the distances to various galaxies. The first thing discovered was that our galaxy is surrounded by about 20 other galaxies (including the Andromeda Galaxy) most of which are much smaller. This group of neighboring galaxies is called the **local group**. Venturing farther into the Universe, there are no more galaxies for quite some distance.

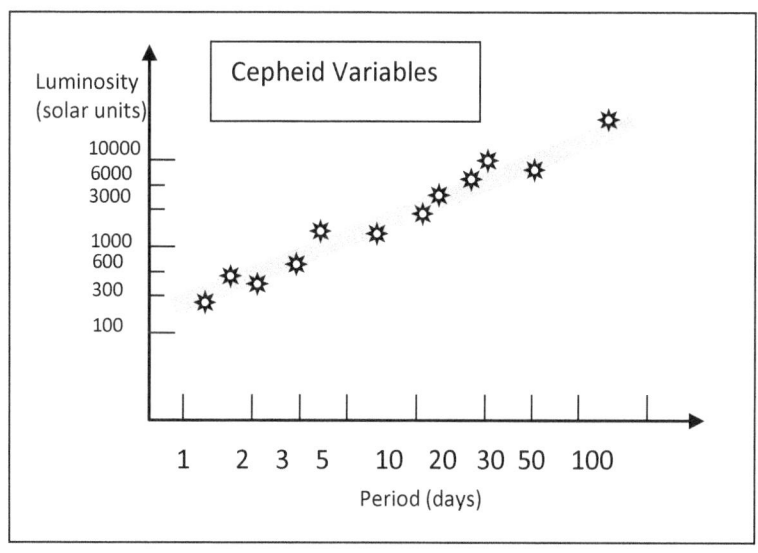

The next moderately rich grouping of galaxies is called the Virgo Cluster—a very large cluster of thousands of galaxies. And beyond the Virgo Cluster there are a number of other groups of galaxies. These neighboring clusters of galaxies compose what is known as the **local supercluster**. Upon further investigation, it now appears that most of the Universe is composed of clusters of galactic superclusters separated by vast regions of emptiness. The overarching structure of the Universe appears to be small regions of material separated by largely empty regions.

7.4 Red Shift

Taking light from galaxies and separating it into a spectrum reveals some very interesting aspects of galaxies. In addition to measuring the average temperature of the stars in a galaxy (hot and young stars produce a blue color if enough are present) and the chemical composition (like stars, galaxies are mostly hydrogen), a third secret can be deduced from a spectrum. Because the light from a receding galaxy is subject to a phenomenon called the Doppler effect, we can use the apparent shifts in wavelength of characteristic absorption lines—called the red shift—to measure the recessional speed a galaxy.

We have all heard what happens to the pitch of an ambulance siren or a racecar as it passes us—it decreases. This effect, whether we are talking about sound or light, is called the Doppler effect. To help you better understand the source of this effect, consider the following analogy: Pat is sitting in the back of a pickup truck with a bushel basket of potatoes that she is throwing out the back at a regular rate (say, one every second). If Chris is sitting just behind the truck, he will of course get hit by the potatoes at the same rate—once every second. But, what happens if the truck starts to drive away? Each time a new potato is thrown, the truck will have moved farther away making the distance to Chris increase. Assuming that the rate at which the potatoes are thrown does not change, the rate at which they reach Chris will actually go down (i.e., they will arrive at intervals of greater than one second). If the truck then stops and begins backing up towards Chris, the reverse effect will occur—the potatoes will start arriving at intervals of less than one second. So, if Chris had his back turned and his eyes closed (wouldn't you?) then, just by judging the rate at which he was being hit, he could tell whether the truck was driving towards or away from him.

This situation is analogous to the case of the ambulance siren. The sound produced by the siren is just a pulsating disturbance (about 500 to 1000 times per second) that travels at a constant rate through the air. You can almost think of the siren as "throwing" sound waves at you. When the siren is approaching, the pulses arrive just a little more frequently than they are emitted (the pitch goes up) whereas when the siren is moving away (after it passes you), the pulses arrive at a slightly lower frequency than they are emitted (the pitch goes down).

> **Box 7-2: Unreasonable or Imprudent?**
>
> Many of us have had the experience of being stopped for exceeding the speed limit and the conversation between you and the officer is always the same.
>
> > **Officer:** Do you know how fast you were going?
> > **Driver:** Well, I'm not exactly sure.
> > **Officer:** Well I am!!
>
> Now how does the officer know? Some people mistakenly think that a radar gun sends out a beam of light traveling at a predetermined speed and the time it takes for the light to go out, bounce of the speeding car, and return provides a measure of the car's speed. If, however, you recognize that light travels enormously fast, about 186,000 miles per second, you realize that the travel time is far too short to be measured easily. Instead, radar speed guns work in the following way. A light wave with a known wavelength is projected at a speeding car. When the light wave bounces off of the car, its wavelength changes. If the distance between the gun and the car is decreasing, the wavelength is decreased—called a blue shift. By measuring exactly how much the wavelength has changed, it is possible to calculate the rate at which the distance between the gun and the car is closing. Likewise, if the emitted light wave bounces of a car when the distance between the car and the gun is increasing, the wavelength becomes longer—it is red shifted. It is this change in wavelength that provides the rate of change of distance between the gun and the car.

In the Unit on the nature of light, we learned that light, like sound, has a wave-like nature. Although there are problems with the analogy, you can think of light being like a sound wave except that instead of talking about pitch we talk about color. If a light source is receding, the wavelength that we receive is actually longer than if the source were stationary; we say that the light has been red-shifted. If a light source is approaching, the wavelength that we receive is shorter than if the source were stationary and we say that the light has been blue-shifted. In fact, Doppler believed that this effect accounted for the different colors of stars we see in the sky. He thought that stars that appeared quite blue were approaching us whereas reddish stars were receding. Of course, we now know that this color difference results from the different temperatures of stars.

The difficulty faced by Doppler emphasizes an important point. You can only learn about the motion of an object based on its color if you know what it should look like at rest. The key to using the Doppler shift to measure the motion of distant galaxies is the use of specific absorption lines that can be identified in the spectra of all galaxies. As we learned in Unit 4, the exact wavelengths of these absorption lines depend only on the specific chemical elements responsible and therefore would be constant if all galaxies were at rest with respect to us. The absorption lines from distant galaxies do indeed show the same familiar patterns that can be reproduced in Earth-bound laboratories. But as early twentieth century astronomers discovered, the patterns are consistently shifted to longer wavelengths—the spectra have been red-shifted. By measuring the exact amount of the shift, the recessional velocities of the galaxies can be calculated. The percentage of shift is called the "z factor" and is largest for the galaxies moving the fastest.

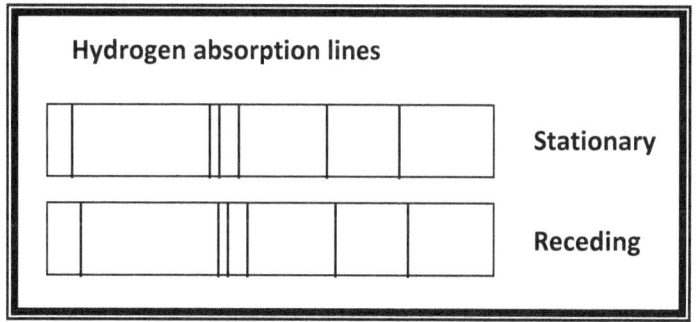

As Hubble began to compare the distances to galaxies to their corresponding velocities, he noticed a startling and consistent pattern. First, he noted that almost every galaxy he observed had its spectrum red-shifted to longer wavelengths. If all galaxies are red shifted, then all galaxies must be moving away from us. Moreover, he discovered that the farther away a galaxy is, the faster it is moving away from us. This direct relationship between a galaxy's distance and its recessional velocity is known as **Hubble's Law**.

Hubble's Law has profound implications. The most striking result is that if all galaxies are moving apart, then at some time in the past the galaxies must have been much closer together. And if galaxies were much closer together in the past and they are not close today, then there must have been some incredible and awesome event to make the distances between galaxies increase dramatically. Astronomers call this extraordinary event the **Big Bang**.

There are some seemingly strange issues that immediately arise when talking about the proverbial Big Bang. The first evident issue is: if galaxies in all directions seem to be going away from the Milky Way, does that mean that the Milky Way is the center of the Universe and, accordingly, the location of the Big Bang? It turns out that in any expanding system of objects, the center of expansion is very difficult to define. Consider a seemingly unrelated example. When a world-class chef prepares raisin-bread, she starts with a lump of dough that has raisins scattered throughout. When the dough goes into the oven, the dough rises and at some point becomes raisin-bread. If there were a small, imaginary camera on one of the raisins, it would record that every other raisin moves away from it as the bread expands. This is true regardless of whether the reference raisin was in the middle of the loaf or near the edge. The raisins that started out nearby would move away uniformly in all directions and those that started out farther away would appear to move a greater amount. The same situation exists in an expanding Universe—no matter where you stand, it always looks like everything is moving away equally in all directions. This is because no matter where you start, you are moving along with the expansion.

Even more odd, this situation encourages us to ask the question of what kind of empty void these far-flung galaxies are expanding into? It turns out that galaxies are not flying apart at all—just like with the raisin bread, it is actually the space between the galaxies that is expanding. There is no "void" that galaxies are moving into. The space between them is growing and therefore their separation is increasing.

Hubble's law of expansion—the farther away a galaxy is, the faster it is receding—can be represented by the simple equation **v=HD**. This equation tells us that the speed at which a galaxy is receding (v) can be found by multiplying its distance (D) by a value (H) that tells us about how fast the Universe is expanding. The constant H is called the **Hubble Constant**, which, because of the difficulty in accurately measuring the distances to far-off galaxies, is remarkably difficult to determine accurately. Current estimates for the Hubble constant generally range from 60 to 75 km/s/Mpc. What this means is that a galaxy at a distance of 1 million parsecs (1 Mpc) is receding from us at somewhere between 60 and 75 kilometers per second. Galaxies farther than this are receding faster while those closer by are moving away at a more moderate rate.

The value of the Hubble constant is a very important number—it says something about the evolution of the Universe. As we look out into the Universe, we can say two things about galaxies: (1) how far apart they are and (2) how fast they are moving apart. This means that, if we run time backwards, we can determine how long it took the galaxies to reach their present location and, therefore, determine the age of the Universe. Estimated values of the Hubble Constant imply that the Universe is somewhere between 10 and 16 billion years old. This age is subject to some argument, not because the recessional velocities are unknown, but because the distance to galaxies is difficult to measure accurately. Uncertainties in measuring distance lead to uncertainties in the Hubble constant and therefore to uncertainties in estimates of the age of the Universe. One of the primary goals for the Hubble Space Telescope (HST) is to look for distance indicators in faint galaxies to more accurately determine the Hubble constant and, subsequently, the age of the Universe.

7.5 Quasars

Possibly the most distant and faint objects in the Universe are known as quasi-stellar objects, or **quasars** for short. Not discovered until the 1960s, these objects are only visible by looking at long wavelength, low energy radio waves that usually originate from very cool or dim objects. When astronomers first began to examine the spectra from these star-like objects, they were completely baffled. The spectra looked unlike anything they had ever seen. Eventually, astronomers recognized that the spectra from these objects was indeed familiar to them, it was just Doppler-shifted so far toward the red that no one had noticed what these objects were. The enormous red shift means that these objects have unimaginably large recessional velocities and, when plotted on the Hubble Diagram of distance versus recessional velocity, seem to be at distances approaching 15 billion light years! If the bright Andromeda Galaxy were at a distance of 15 billion light years, there is no possible way that modern telescopes could observe it—the galaxy would just be too dim to see. This implies that Quasars must be incredibly bright with luminosities of hundreds or even thousands of times that of the entire Milky Way Galaxy. And, more impressively, the source of this enormous amount of energy is concentrated in a space smaller than the distance from our Sun to its nearest neighbor star. Astronomers still do not fully understand the nature of quasars. When we see quasars at a distance of 15 billion light years, we are looking back in time to see not what they look like today, but rather as they appeared as many as 15 billion years ago, around the time of the Big Bang. Maybe we are seeing infant galaxies just beginning to form. Maybe we are seeing early Universe objects that no longer exist. It is a mystery that contemporary astronomers are aggressively trying to solve.

7.6 Epilogue: Cosmology

Cosmology is the study of the origin and evolution of the Universe. Unlike astrophysicists—who can study billions of different stars of every possible size, composition, and age—cosmologists have only one universe to study. Moreover, they can only study our Universe at one particular instant in its evolution—the present. These fundamental constraints make it all the more amazing that a single theoretical framework has emerged as the dominant theory for explaining our Universe. This theory is called the **Big Bang model**. It is certainly true that cosmologists actively debate the details of the Big Bang model but this should come as no surprise given the complexity of the current Universe it is attempting to explain. What is more important is that this single complex theory brings together ideas ranging from the smallest scales of high-energy particle physics to the largest scales of the structure of the Universe into one reasonably consistent picture. The Big Bang model not only predicts the current expansion of the Universe but, by detailing the interaction of particles and light from the very time that the Universe came into existence, it also makes specific predictions that are subject to testing. For instance, the Big Bang theory predicts the fusing of hydrogen into helium in the universe's early history predicting an abundance of helium in good agreement with what is observed today. The helium produced in stars only accounts for about 10% of this abundance.

Perhaps the greatest triumph of the Big Bang model came in its prediction and the subsequent verification of a remnant background radiation that traces its history back to a time when the Universe was a mere 300,000 years old at an average temperature of 3000 degrees. At that time, radiation filled all space and had the familiar blackbody radiation spectrum that we saw in the unit on the nature of light. As the Universe has expanded, this radiation has red-shifted to longer and longer wavelengths, which corresponds to cooler and cooler temperatures. Launched in 1990, NASA's Cosmic Background Explorer Satellite (COBE) accurately measured the cosmic background radiation and found a perfect blackbody spectrum consistent with a temperature of only 2.7K (that is, 2.7 degrees above absolute zero). This observation serves as one of many confirmations of the theory.

Observation	Inference
Almost all galaxies are red-shifted	The distances between our galaxy and most others are increasing
The most distant galaxies exhibit the greatest red-shift	The space between all galaxies is expanding
The ratio of recessional velocity to distance is between 60 and 75 km/s per megaparsec and is called the Hubble Constant	The Universe has been expanding for 10 to 16 billion years
The Cosmic Background Explorer (COBE) found that the temperature of intergalactic space was not zero	The Universe has not yet cooled from the rapid Big Bang expansion

One of the most interesting questions, whose answer is not predicted by the Big Bang model, concerns the ultimate fate of the Universe. Will the Universe continue to expand forever or does it have enough mass to eventually halt the expansion and collapse again in an event that has come to be known as the **Big Crunch**? Although still open to debate, the preponderance of evidence seems to suggest that the average density of material in the Universe is sub-critical meaning that the Universe will continue to expand forever.

The Big Bang is actually a lousy name for this fantastic event. The words "Big Bang" bring forth images of explosions and fire—certainly not likely in the early Universe. In fact, in the 1980s, the Planetary Society held a contest to see if there was a better name for the Big Bang. Kids and adults all over the world sent in entries with suggestions for a new and better name. In the end the judges decided that, for now, the name Big Bang would stay.

Chapter 8
Measuring the Sky

One aspect of astronomy is called 'modern astronomy'. However, astronomy has been around for a very long time! We can make measurements of the apparent shifts in a nearby star's position to deduce distance. Exploring the properties of the light from a star teaches us about temperature, composition, and motion. Common characteristics and trends are used by astronomers to extend the distance ladder beyond the limits of trigonometric parallax. This story, however, ignores many of the controversies and competing theories that are an integral part of scientific progress in order to give an overview of the scope of the topic. The very technical and mathematical nature of modern astronomy, as well as the enormous volume of modern astronomical knowledge, can make these professional debates difficult to follow.

In contrast, lively debates about the motions of the planets or the Earth's position within the Solar System provide us with a unique opportunity to study the evolution of scientific ideas without the need for an advanced level of mathematical understanding. In studying the changes in astronomical models from the time of the ancient Greeks through to the recent discovery of the outer planets Uranus, Neptune, and Pluto, we will investigate how models have developed, evolved, and finally been abandoned in favor of better models. We will see how both observational evidence and entrenched philosophical beliefs have influenced how different generations of scientists and astronomers have come to understand the cosmos and our place in it. The goal of this unit is not to discredit those early investigators, rather, it is to highlight how scientific theories represent the best possible understanding at the time. When challenged by new and contradictory observations, theories can be either refined or replaced by new ideas. This perspective should be remembered even when evaluating modern scientific theories.

8.1 Models, Theories, and Laws

Astronomers often use the terms **model**, **theory**, and **law**. Before proceeding, it is worth clarifying how these terms are used—definitions and interpretations that are by no means uniformly agreed upon by the entire scientific community. The most difficult problem is to attempt to define these terms without some reference to the idea of objective "truth," which is itself a delicate topic.

If the first rule of science is that all knowledge is open to revision based on new experiments or observations then the second rule is that scientists use models. **Models** are used as an aid to understanding, to make predictions, to conduct experiments, and to simplify complex phenomena. Scientists use two flavors of models: scale models and conceptual models. Scale models are smaller or larger versions of a physical object. For example, a small model airplane can be placed in a wind tunnel to study the aerodynamic behavior of the wings. The principle use of such a model is to test predictions about how well the wings are designed. This is the type of model that most people are familiar with. Often more useful to scientists, however, are conceptual models, which are mental abstractions designed to capture the critical features of the system under study.

To be useful, a model need not represent the entire inner truth about a system. It is enough that the model reflects only the aspects of the system that are central to the problem under study. For instance, when calculating the forces on a girder used in building a bridge, it is perfectly reasonable to model the girder as a continuous and solid piece of metal even though we understand that, at a more fundamental level, the girder is really a collection of individual atoms. As long as we do not use our simplified model to try to understand the girder at scales on the order of 10^{-10}m (at which scale the fact that it is indeed a collection of individual atoms begins to matter) we are quite safe in using it. Our belief in the "reality" or "truth" of the model may change with time but, just because we no longer believe the model to be complete does not mean that it is no longer useful.

There are many models of the heavens that at one point were thought to represent the truth about the Universe but have since been shown to be incompatible with our growing body of knowledge. These models were generated to explain the then available data and today, even though we recognize shortcomings, these models are useful in organizing our thoughts about the specific observations that they were first designed to explain. For example, even though the modern view puts the Sun at the center of the Solar System and the Earth in orbit about it—called a **heliocentric** model—the changing position of the Sun in the sky is probably more easily conceptualized by a rising and setting model in which the Sun moves about a stationary Earth, called a **geocentric** model.

Many scientists believe that we can never really know the "truth" about our physical world—our models will always be simplifications of the truth. A **theory** consists of both a model and a statement as the model's relative truth. When the heliocentric theory of the Solar System was first advanced, it was a proposed that the motions of the planets could be more easily understood if the Sun were considered to be at the center of the Solar System. Moreover, it was also a statement that the Sun really is at the center of the Solar System. Thus it was the geocentric (Earth-centered) theory of "truth" that was under attack, not the geocentric model. The geocentric perspective remains incredibly useful in organizing our thoughts about the rising and setting cycles of the Sun.

A scientific **law** is probably best thought of as a theory that has attained the status of a law when repeatedly confirmed by experiments and is found to be a necessary component of a wide range of other theories. For instance, there have been a number of competing theories advanced to explain the evolution of our own Solar System. These theories differed in many ways but they shared certain features such as abiding Newton's law of universal gravitation. A hallmark of science is that even laws may eventually be proven wrong but, just as with models, this does not necessarily mean they are no longer useful. Newton's law of gravitation, the foundation of so much of modern astronomy, has been shown to be incomplete and is no longer considered to be the best theory of gravity. We now believe that Einstein's theory of general relativity—necessary to understand the effect of very massive bodies—provides the best available description of gravity. Yet we still make use of Newton's law of gravity because of its wide domain of applicability (it works for virtually all "everyday" problems), and it retains its status as a fundamental law of physics.

8.2 The Beginning of Astronomy

It is impossible to trace the roots of astronomy to one unique starting point. Moreover, different cultures, often in isolation, had very sophisticated astronomical ideas. Although the scattered remains of the astronomical knowledge of ancient civilizations is highly fragmented, it is clear that many civilizations, ranging from the Druids of the British Isles to the Aborigines of Australia, developed sophisticated understandings of the heavens both for religious and calendar keeping purposes. For instance, the Mayans placed a lot of significance in eclipses, feeling that they foretold evil. The records of this magnificent society that survived the devastating Spanish conquest reveal tables of eclipses that not only recorded their past occurrence but also predicted those that had yet to occur.

It was likely the need for accurate calendars that motivated the first careful observations of the motions in the heavens. The ability to identify the appropriate timing for the planting or harvest time was clearly important to the earliest societies. As societies became more organized as municipal institutions, the need for accurate calendars accelerated. Along the Euphrates River, about 70 miles from the current city of Baghdad, was the ancient city of Babylon. The Babylonians developed an advanced arithmetic that forms an important part of the history of mathematics. In fact, the modern division of the hour into sixty minutes and the minute into sixty seconds can be traced to the Babylonians. Given the tool of an advanced mathematics, the Babylonians were also able to develop a complex understanding of motions of stars and planets. Their interest in this extended beyond the purely academic because they believed that particular events in the sky foretold significant events on Earth. Observational records from China, Japan and Korea also go back several thousands of years and include early records of supernova events.

Many cultures have made important astronomical discoveries and the fact that they are omitted in this unit is not meant to diminish their importance. In this early astronomy unit, we focus almost exclusively Mediterranean astronomy as this is certainly the work that most influenced great thinkers in the following centuries. Further, the Greeks moved beyond observation and accurate record keeping to develop models that could explain their observations—they turned astronomy into a true science.

The Greek civilization, beginning in the seventh century BC, is where the historical record actually allows us to associate powerful ideas with specific individuals. This society of astronomer-philosophers were asking an entirely new set of questions compared to the astronomer-priests (whose only interest was in interpreting stellar motions). The detailed story of the development of Greek philosophy and astronomy is indeed interesting, but lies beyond the scope of this text. Our goal is to see how the Greek astronomers constructed a robust model of heavenly motion that ultimately endured for more than a millennium from *observational evidence*.

8.3 Early Understanding of the Earth

At the time Columbus sailed the seas in 1492 AD, the size and shape of the Earth was well understood by scholars, if not by most commoners. 1500 years earlier, the Greeks were not only aware of the Earth's spherical shape but also had a reasonable measurement of the size of the Earth that was remarkably accurate.

So how do we know that the Earth is round? There are a number of observations to which the Greek's could have appealed to reach this conclusion:

- As a boat disappears into the distance it "falls" out of sight well before it is out of true visual range. In fact, if you were watching from a second boat you could simply climb up the mast and bring the first boat back into view. This is easily explained in terms of a curved Earth instead of a flat surface.

- During a lunar eclipse, the shadow cast by the Earth on the Moon is round (although only part of it cuts across the Moon at any given time). A little consideration should convince you that the *only* shape that will always cast a circular shadow is a sphere. **Aristotle** (384 BC – 322 BC) used this evidence to support his claim of a spherical Earth.

- As you travel from north to south you lose sight of some stars while others come into view that can never be seen in the more northerly latitudes. A curved surface easily explains this observation.

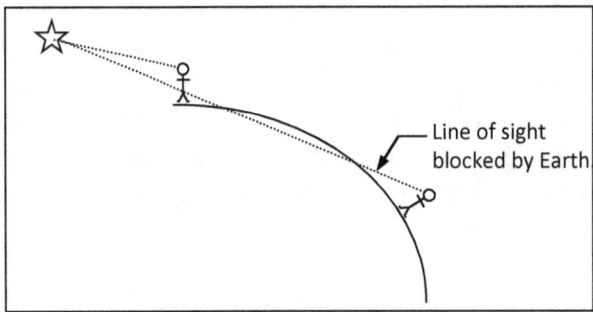

Once satisfied that the Earth is indeed a sphere—or nearly so—the next question that arises is one of size. Determining the Earth's size is clearly not amenable to techniques of casual observation. Actually, it is amazing to think that **Eratosthenes** (276 BC – 195 BC) first accomplished this more than 2000 years ago. Using simple instruments and a little traveling, he was able to produce a remarkably accurate estimate of the circumference of the Earth. His method was based on the observation that, even when at its highest point in the sky, the Sun always cast at least a short shadow in his home of Alexandria. But, according to some reports, the Sun would cast no shadow (in fact, it would shine straight down a deep well) at noon on the day of the summer solstice in the more southerly city of Syene—now known as Aswan.

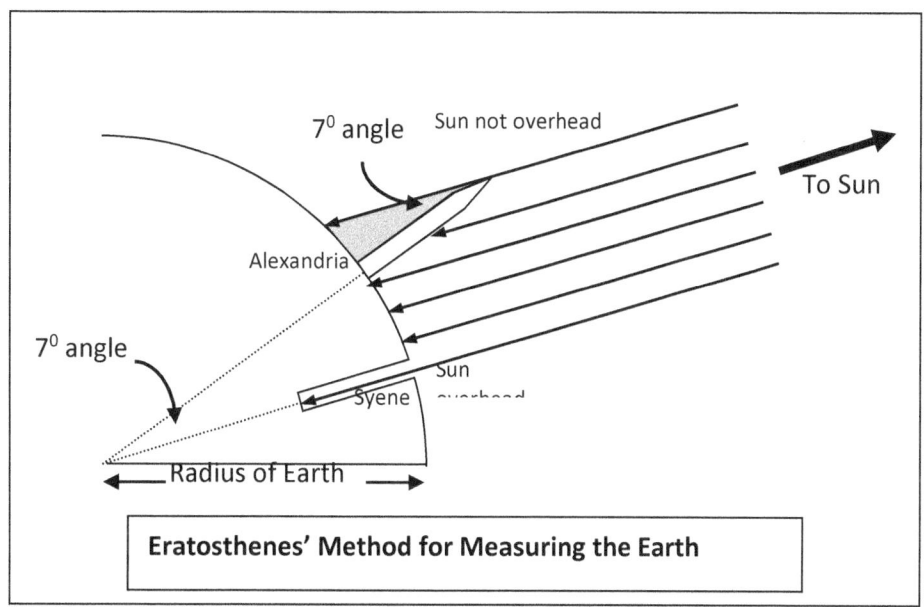

Eratosthenes' Method for Measuring the Earth

Undoubtedly, many people were aware of this phenomenon but it was Eratosthenes who recognized how it could be used to determine the size of the Earth. Under the assumption that the Sun was very far away, Eratosthenes recognized that the angle of the shadow cast by a tall stick in Alexandria at noon on the summer solstice was the same as the angle between Alexandria and Syene as viewed from the center of the Earth (as shown in the figure above). Eratosthenes measured the angle of the shortest shadow to be about 7^0. This means that the distance from Alexandria to Syene was seven three-hundred-and-sixtieths (7/360) of the distance all the way around the spherical Earth. The legend is that Eratosthenes paid someone to pace off the distance between the two cities (nearly 500 miles!) so that he could complete his calculation. The conclusion he reached was that the Earth's circumference was about 25, 000 miles—remarkably close to what we know it to be today.

8.4   The Moon and the Sun

There are two good reasons for the Greeks to have concluded that the Moon is in fact closer to Earth than the Sun:

- It was recognized that during a solar eclipse, the Moon moves in front of the Sun. To cast its shadow on Earth, the Moon must be closer.

- The Moon appears to moves faster against the background stars than the Sun does. It takes a little less than a month for the Moon to complete its path relative to the background stars whereas the Sun takes a full year. Consistent with everyday experience, you probably notice that things that are far away appear to move more slowly. For example, compare the apparent speed of car passing close to you on the street to the apparent speed of a distant aircraft flying at 35,000 feet.

The relative size of the Moon seems to have been first determined by the astronomer **Aristarchus of Samos** (ca. 310-230 BC) about 75 years prior to Eratosthenes' measurement of the size of the Earth. Aristarchus made many important contributions to astronomy but is certainly best remembered as the first scientist to suggest that the Sun, and not the Earth, is at the center of the universe (an idea that is often inappropriately credited to Copernicus, who reached this conclusion some 1,700 years later). Aristarchus based this conclusion on his measurements of the relative sizes of the celestial bodies. He showed the Moon to be about one-third the size of the Earth and the Sun to be about twenty times farther away from the Earth than the Moon. If both of these conclusions were true, it followed that the Sun must be about twenty times larger than the Moon and therefore at least seven times larger than the Earth. Accepting this conclusion, he found it hard to imagine that the large Sun could possibly be in orbit about a smaller Earth.

Aristarchus based his estimate of the relative sizes of the Earth and Moon on the size of the Earth's shadow cast on the Moon during a lunar eclipse. Aristotle had earlier used this same round shadow to support his theory that the Earth was spherical. Using this method, Aristarchus concluded that the Earth's radius was about three times that of the Moon, which was reasonably close to the modern value of 3.7.

Aristarchus' determination of the relative distances to the Sun and the Moon is best understood by examining the figure at the top of the opposite page. His approach relied on the observation that the first quarter Moon must form one corner of a right angle triangle, with the Sun and Earth at the other two corners. Thus, by measuring only the angle between the direction to the Moon and the direction to the Sun (called (θ in the figure,) he concluded that the Sun was 20 times farther from the Earth than the Moon.

In principle there is nothing wrong with this method but Aristarchus' estimate was about 20 times too low—the Sun is actually more than 400 times farther away from the Earth than the Moon. His difficulty was that the Sun-Moon angle θ is very close to 90 degrees because of the enormous distance to the Sun (not to scale in the figure). This makes it very hard to measure precisely. Aristarchus found a value for the angle of 87^0 whereas it is actually about 89.8^0. This might seem to be a small error (less than 3 degrees), but unfortunately this method is very sensitive to small errors in measurement, resulting in large errors in the predictions.

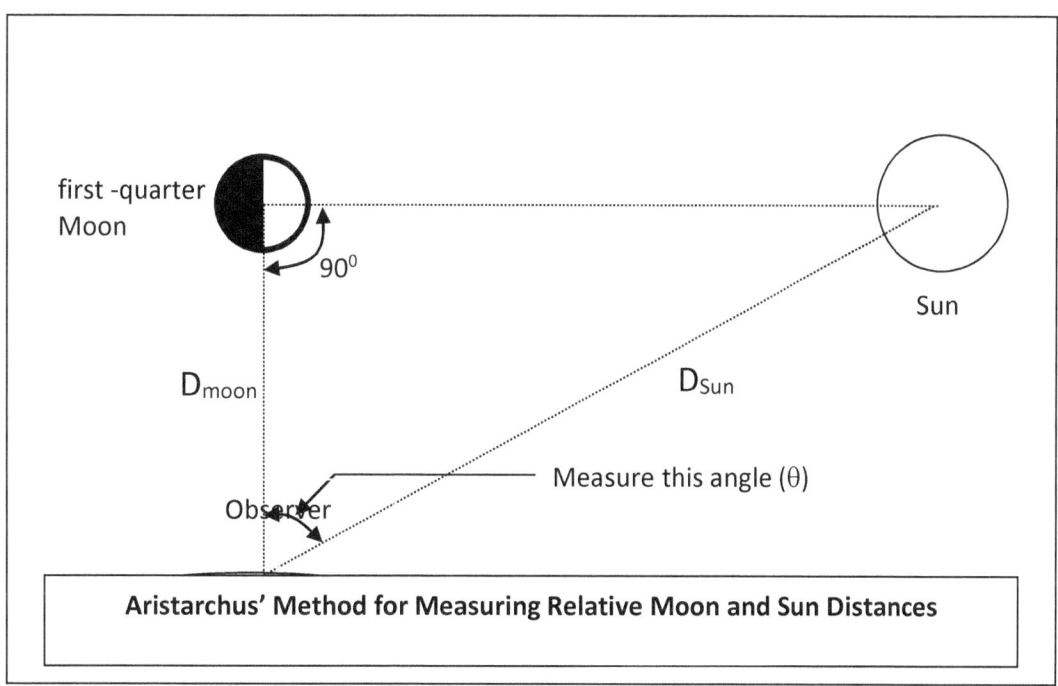

Aristarchus' Method for Measuring Relative Moon and Sun Distances

Armed with his calculation of the relative distances to the Sun and Moon, as well as the relative sizes of the Moon and Earth, Aristarchus proceeded to estimate the relative size of the Sun. To understand his procedure requires two observations:

- The Moon and Sun appear to be about the same size in the sky. It follows that if the Sun is farther away it must be bigger.

- If two objects appear to be the same size and one happens to be twice as far away, it must be twice a big because the size that something appears is directly proportional to its distance. Accordingly, Aristarchus believed the Sun to be 20 times farther away than the Moon and, since they appeared the same size in the sky, he concluded that the Sun must be about 20 times larger than the Moon.

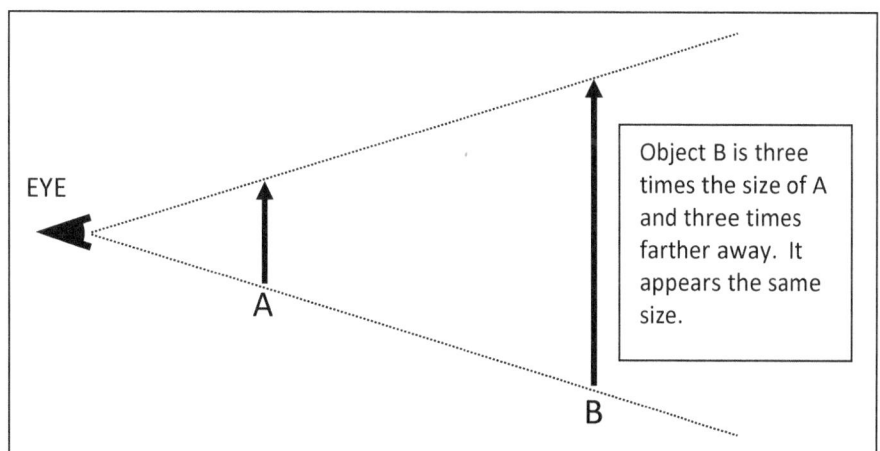

Continuing with this logic, since the Moon was about one-third the size of the Earth he concluded that the Sun must be about 7 times larger than the Earth. Certainly his numbers are vastly different than today's values, but he did get the relative order correct. His error can be traced to the substantial difficulty in accurately determining the Sun-Moon angle required for determining the distance to the Sun.

Chapter 9
Greek Models of the Heavens

The human fascination with the night sky can be traced virtually to the beginning of recorded history, and yet it has been for less than 400 years that this curiosity has been informed by the magnified images of the telescope. For many thousands of years prior to Galileo's first telescope observations, those interested in unraveling the mysteries of the sky were limited to what they could see with the unaided eye—the naked-eye observables. These observables consisted of two basic types: what objects looked like and how they moved. Of course, most objects in the sky just look like points of light and therefore interest in them was limited to understanding their motions. Only for a few objects—the Sun, the Moon, and the Earth itself—could anyone even talk about such properties as size or distance.

9.1 The Pythagorean Model

In addition to pursuing questions of size and distance, Greek astronomers began to explore how to explain the motions of the heavens. The first record of a model to explain these motions comes from a group of philosophers known as the Pythagoreans (ca. 550 BC). You are likely familiar with this name because of a rather famous theorem in geometry relating the sides of a right triangle. However, the interests of this group extended well beyond the field of geometry. Their interests also included mathematics, music, and astronomy. It is in astronomy that they made monumental contributions, both philosophical and technical, that would endure for more than two millennia! To understand the **Pythagorean Model** of the heavens, we need to review exactly what observations the Pythagoreans were attempting to account for and, because the Pythagoreans believed in a geocentric universe with a stationary Earth, we will consider it from the Earth-bound perspective. This is a universe in which the stars appear to move in circular paths across the sky. For stars near the stationary North Star—called circumpolar stars—we observe a complete circle. Stars that are more southerly, called seasonal stars, move in a circle that is interrupted by the Earth's horizon such that we only observe a portion of the circle (see section 2.3).

In addition to stars rising and setting and rising again every 23 hours and 56 minutes, the other important celestial body whose motions would have to be explained by any model is the Sun. The Sun, like the stars, rises in the East and sets in the West. However, the Sun moves around the Earth about 4 minutes per day slower than the stars—it takes the Sun 24 hours to rise and set and rise again. Both of these observations must be accounted for in a working model of the Universe.

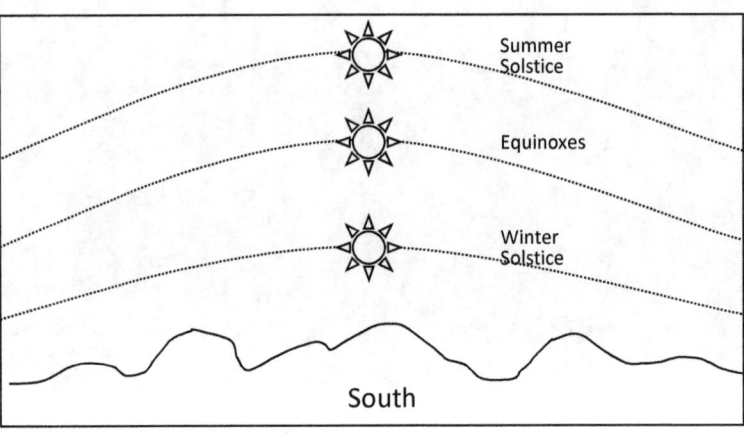

The path that the Sun follows across the sky varies with the seasons. At the winter solstice, about December 21, the Sun tracks along its lowest path in the sky, resulting in very short days and long nights. At the time of both the March (vernal) and September (autumnal) equinoxes, about March and September 21 respectively, the Sun rises exactly in the East and sets exactly in the West and is accompanied by about 12 hours of daylight and 12 of darkness. On the summer solstice, near June 21, the Sun tracks its highest path across the sky leading to long days and brief nights. Notice that, due to the tilt of the planet, the seasons are reversed in the southern hemisphere.

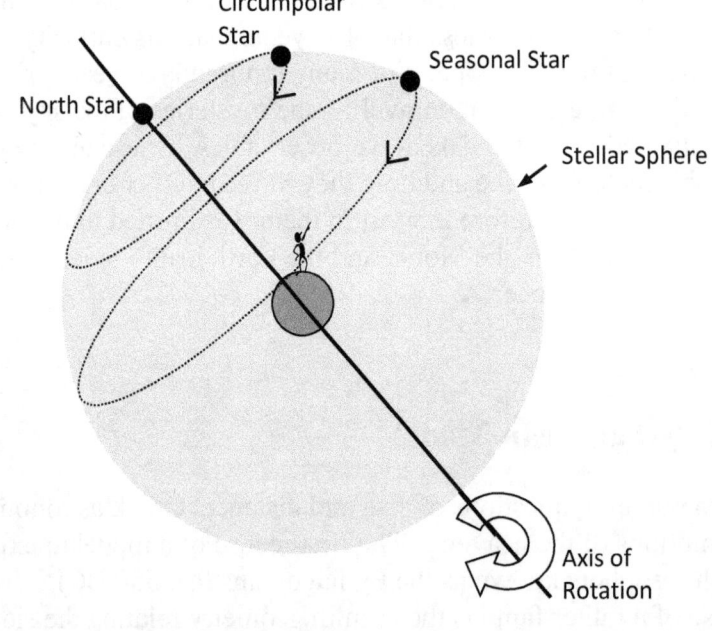

The most important element in the Pythagorean model is the **stellar sphere**. The Pythagoreans envisioned the Earth surrounded by a giant rotating sphere "painted" with stars. All stars were an equal distance from the Earth and their motions were simply the result of the rotation of the stellar sphere about its north-south axis. To account for the period of stellar motion, the stellar sphere had to rotate once about its axis every 23 hours, 56 minutes. It accounts very simply for the motions of the stars and, in fact, is the simplest way to think about the night sky when operating a telescope.

To account for the motion of the Sun within such a scheme requires some additional complications. If the Sun moved at exactly the same rate as the stars then the Sun could be imagined as fastened directly to the stellar sphere. However, because the Sun moves slightly slower than the stars, it changes its position on the stellar sphere throughout the year. The path the Sun follows relative to the background stars is called the **ecliptic**, imagined by the Pythagoreans as a track around the stellar sphere on which the Sun moved. Each day the Sun's position on the stellar sphere changed slightly—it slipped backwards or eastward a bit—so that each night it would set next to a different star. After one year the Sun would get "lapped" by the stellar sphere and return to the same position it had occupied a year earlier. To account for the changing path of the Sun with the seasons, it was necessary to "tip" the ecliptic relative to the stellar sphere's equator. In this way, for part of the year the Sun would be a long way from the North Star and would therefore travel along a path low in the sky. For the other the other half of the year, the Sun would be closer to the North Star and take a higher path in the sky.

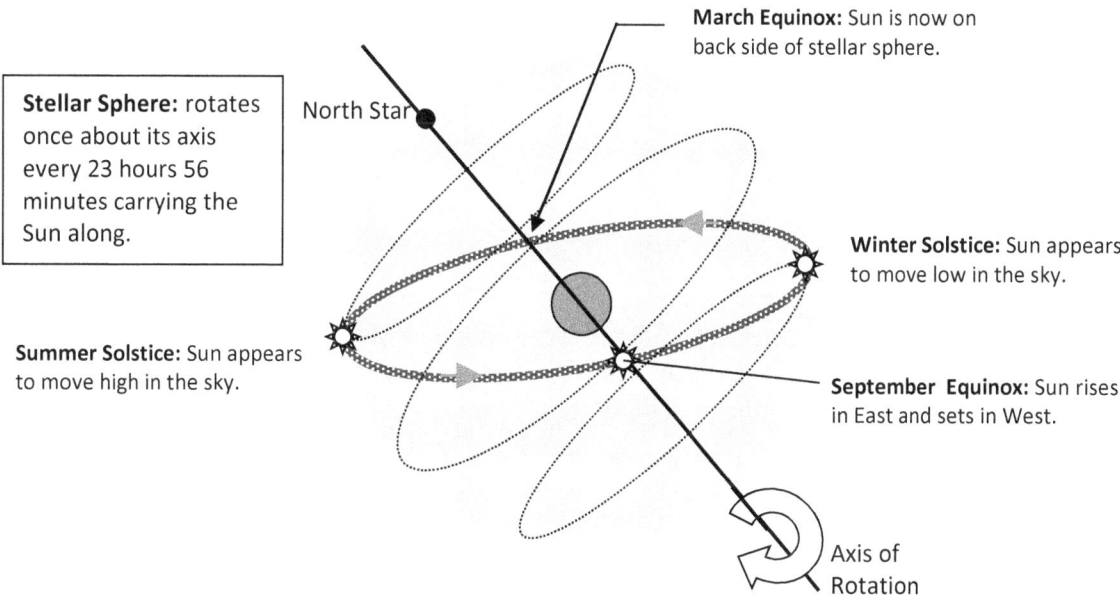

The Pythagorean model accounts remarkably well for the constant motion of the thousands of stars visible in the night sky. The stellar sphere model also accurately describes the daily and seasonal motions of the Sun. Although we now recognize that this geocentric theory fails to capture much of the "truth" about our Sun-centered Solar System, it still provides a productive way to conceptualize the Sun and stars to understand their basic motions.

As well as offering a reasonable model to explain the observed motions of the heavens, the Pythogoreans also set forth some philosophical constraints that continued to influence the thinking of astronomers for the next two thousand years, well after the basic Pythagorean model was long forgotten,. The Pythogoreans appear to have first proposed an idea that was later championed by the great philosopher **Aristotle**. They proposed a sharp distinction between the **terrestrial** and **celestial** domains; the Earthly domain was imperfect and changeable whereas the celestial domain was seen as the perfect manifestation of God. The heavens, unlike the Earth below, were thought to represent perfection and were therefore **immutable** (unchanging). The Moon was thought to separate the imperfect terrestrial domain from the perfect celestial domain and was thus seen as a perfect jewel in the sky. Such events as lightening, meteors, and comets were temporary and therefore could <u>not</u>

exist in the perfect celestial domain—these all had to be atmospheric effects and part of the terrestrial domain.

The second important philosophical constraint to be introduced by the Pythagoreans was the notion that perfect circles must govern all heavenly motions. They were strong believers in the power of numbers and geometry, believing that the circle represented the ideal form. It was therefore natural to extend this belief and assume that the motion of the objects in the perfect celestial domain should strictly follow paths governed by the circle.

It is hard to underestimate the powerful impact that these ideas had on astronomy. For instance, the Chinese accurately recorded the supernova that resulted in the Crab Nebula in the eleventh century. Yet, European astronomers failed to note this wondrous new and very bright star in the sky. So deeply held was their belief in the immutability of the heavens that they must have dismissed this new "star" as some sort of atmospheric event. As more evidence of the power of the Pythogoreans' ideas, consider that two thousand years later, when Copernicus was willing to demote the Earth to being just a planet, he still required planetary orbits around the Sun to be perfect circles!

9.2 The Wandering Stars

There are five objects in the night sky that look much like stars but move across the sky differently than all the other thousands of stars. These star-like objects—named Mercury, Venus, Mars, Jupiter, and Saturn—were well known to the ancient astronomers. Nobody knows who discovered these wandering objects let alone who named them. Today these wandering objects are known as planets, which is derived from the Greek for "wanderers." Although the basic Pythagorean model could account for the motions of most of the objects visible in the night sky, it was these five wandering planets, and attempts to understand their motions, that would eventually lead to the overthrow of the entire geocentric theory.

Most of the points of light that we see in the night sky maintain their relative positions night after night, month after month, and year after year. However, careful observations of the planets, even over a short time, reveal that they move amongst the "fixed" stars. The normal path of the planets against the stars is from west to east—the same direction as the motion of the Sun along the ecliptic. Thus, if Jupiter were observed near the star Altair one night, it would gradually appear more and more to the east of Altair as the weeks progressed. It is important to remember however that both Jupiter and Altair would still be rising in the east and setting in the west every night. Accounting for this motion within the basic Pythagorean model would be difficult enough but, to complicate matters even more, the planets occasionally reverse directions and exhibit what is called **retrograde motion**. Periodically, the planets appear to slow down, pause, and reverse their easterly direction against the background stars during which they are said to be in retrograde motion. To account for such complex motions within a model that only included uniform circular motions was indeed a challenge. The ancient Greek astronomers met this challenge with the development of increasingly complex models that, by combining many, many circular motions, were able to account for planetary motion with remarkable accuracy. This Greek model of the Universe might, to modern eyes, appear arbitrary and contrived, yet it survived as the dominant theory for nearly 1500 years. However, in the end, it was the lack of aesthetic appeal of this model that led to its eventual replacement by a geocentric model.

9.3 Developing the Ptolemaic Model

> **Box 9-1: Telling the Story**
>
> When people tell the history stories of scientific discovery, they tend to present a rather sequential progression of ideas as if one naturally followed from the other in a continuous chain. Of course the reality is usually significantly more complicated with many more individual contributors than history actually records. In tracing the history of Greek astronomy the problem is compounded by the lack of direct sources. In fact, much of what we know about the Greek philosophers comes from the records kept by the Islamic civilizations after the fall of the Greek and Roman empires. It is important to make this point so that you appreciate that the history being traced out is but a sketch of the true efforts that went into developing what has become known as the Ptolemaic model.

Perhaps the first person to bring significant mathematical prowess to the study of models of planetary motion was **Eudoxes of Cnides** (ca. 408 – 355 BC). This talented mathematician and student of Plato devised a system built upon the Pythagorean model that offered a qualitatively accurate explanation for the complicated motions of the planets, including retrograde motion. Eudoxes' model for planetary motion involved postulating the existence of a secondary system of crystalline spheres to carry the Sun, Moon, and each of the planets. These secondary spheres existed inside the stellar sphere, which only carried the stars around the Earth. When a single sphere could not account for the observed retrograde motion of a planet, Eudoxes added one or more additional nested spheres, one inside the other, rotating at different rates. These "spheres-in-spheres" eventually amounted to 26 in his model. This model is very hard to visualize. Although not quantitatively accurate in its predictions, his model must be appreciated for its sophistication. Eudoxes used these multiple spheres as a conceptual model but did not go as far as to suggest that the spheres were real objects.

Aristotle (384 – 322 BC) extended Eudoxes' model and actually proposed a mechanism to drive such a geocentric universe of spheres. He eventually developed an almost clockwork system with the outer spheres providing the drive for the inner ones, a system that eventually contained 55 separate spheres. However, Aristotle's more important contribution was his reification[1] of Eudoxes' spheres. Whereas Eudoxes appears to have treated the spheres as a useful model for reproducing the observed motions, Aristotle elevated this model to the level of a full-fledged theory in which the spheres were real objects and not just theoretical concepts. Aristotle was so influential that not only did physical spheres become part of the reality of ancient astronomers, but philosophers, poets, and scholars continue to refer to celestial spheres even today.

Despite their obvious sophistication, these initial attempts at understanding planetary motion failed to achieve the desired level of technical agreement with actual observation. In this aspect, these early models clearly came up short. As new generations of astronomers sought to develop models that would improve the quantitative agreement, there was a general relaxation of earlier philosophical constraints. In defending this more pragmatic approach, one astronomer wrote:

> *The astronomer must try his utmost to explain celestial motions by the simplest possible hypothesis; but if he fails to do so, he must choose whatever other hypothesis meets the case.*

---

[1] Reify: to make an abstract idea real or concrete.

One of the first mechanisms introduced to improve the agreement between theory and experiment was the **epicycle**, which was first suggested by Apollonius late in the third century BC. Rather than imagining planets to travel along circles centered on the Earth, Apollonius proposed that each planet actually travels around a small circle called an **epicycle** whose center in turn travels along a circular path called a **deferent**. This concept (illustrated in the figure at the top of the opposite page) provides a mechanism to account for the occasional planetary retrograde loops.

The figure shows that the path of the planet, as observed from Earth, appears to move backwards when the planet is near Earth. This property of the epicycle-deferent model also explains the observation that planets tend to look brighter when they are in the retrograde phase of their motion. By altering the sizes of the deferents and epicycles for each of the planets, as well as the rates at which each rotated about its respective center, astronomers could begin to "fine tune" the model to fit observations.

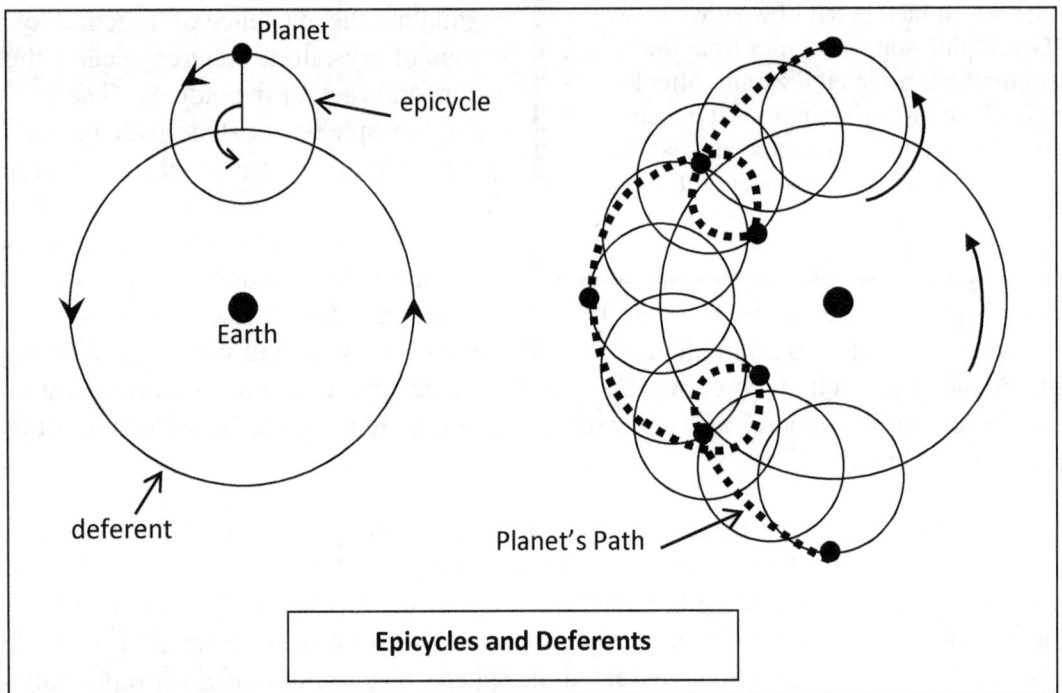

Epicycles and Deferents

Many of the Greek astronomers contributed to refining this basic scheme. Claudius Ptolemaeus of Alexandria (called **Ptolemy**), the last of this group, was the one who crafted the final version in about 140 AD. This model, which contained epicycles and deferents as well as a few new devices, is known simply as the **Ptolemaic Model** and it was to remain the favored theory of planetary motion for nearly 1500 years.

In addition to epicycles and deferents, two novel features in the final version of the Ptolemaic model are worthy of mention. These two features illustrate how astronomers were slowly abandoning much of the fundamental philosophy of Greek astronomy in an attempt to produce agreement with the data. First, in the final Ptolemaic model, the Earth is no longer at the exact center of the system. Second, the deferent does not rotate at a constant rate when observed from Earth. Instead, there is a third point called the **equant,** from which the deferent does appear to rotate at a constant rate. These devices did improve the agreement with the observed motions of the planets, but seem to violate the ideals of a truly geocentric universe in which heavenly bodies move along perfect circles at uniform rates.

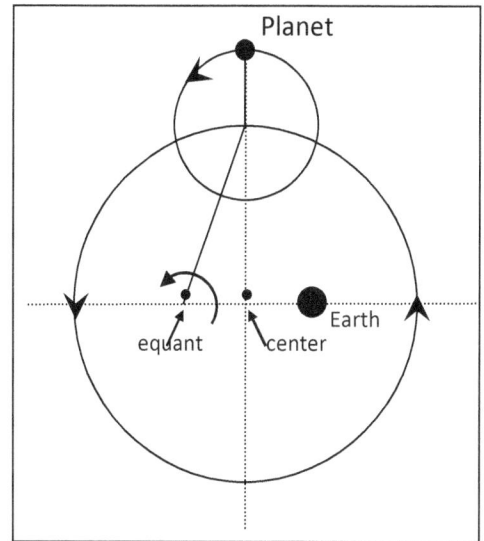

Special mention has to be made about the motions of Venus and Mercury. These planets are often called "evening stars" or "morning stars" because they are only visible near dusk and dawn. That is, they never appear very far from the Sun and therefore are only visible when the Sun is just below the horizon. To address this, Ptolemy made a rather *ad hoc* assumption. He simply insisted that the epicycles of Venus and Mercury do not move independently along their deferents but rather remain fixed along a line joining the Earth to the Sun.

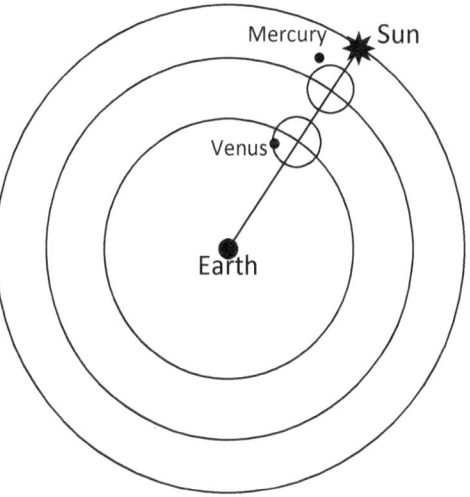

9.4   Epilogue

When we see the simplicity of the Copernican model of the Solar System with the Sun placed at the center, it is hard to understand how it is that the obviously brilliant minds of the Greek civilization could not see clear to abandon their fixation on the geocentric model. We will also see that Copernicus, a revolutionary thinker in so many ways, remained wedded to the notion of circular orbits and eventually had to introduce epicycles into his own model to help correct this defect. On can *never* underestimate the power of ideas and the difficulty in abandoning old ways of thinking, even in the face of compelling evidence.

Chapter 10
A Sun-Centered Universe

The Ptolemaic model described in the last unit is so named because Ptolemy made the most important conceptual contributions to its development. However, the model certainly did not die with Ptolemy in 140 AD. It was kept alive primarily through the work of first Arabian and later, European astronomers. Due to accumulating discrepancies between the model predictions and the observed motions, what they were searching for was the best set of parameters (epicycle and deferent sizes, rotation rates, equant positions, etc.) to bring the model into agreement with observation. King Alfonso X of Castile supported the work of a team of astronomers who, in the middle of the 13th century, completed the last great incarnation of the Ptolemaic model (published as the *Alfonsine Tables*). After centuries of such fiddling, the intellectual climate of the Renaissance was ready for someone to take a bold new step in describing the heavens. This step was taken by Nicolaus Copernicus.

10.1 Nicolaus Copernicus

Born Niklas Koppernigk in 1473, Nicolaus Copernicus was well read in the works of the ancient astronomers and was keenly aware that even the most sophisticated versions of the Ptolemaic system left nagging discrepancies with the observed motions of the planets. However, his search for an alternative conceptualization of the heavenly motions was not fueled so much by these practical considerations as by a fundamental distaste for the complexity of the Ptolemaic model of epicycles.

Copernicus wrote:

> *A system of this sort seemed neither sufficiently absolute nor sufficiently pleasing to the mind.*

More than anything, it was the lack of aesthetic appeal of the Ptolemaic system that led Copernicus on a search for alternatives. Why he first considered the possibility of a heliocentric system—one with the Earth moving about a stationary Sun—is not entirely clear. In his writings, he did allude to the earlier work of at least two Greek philosophers who had entertained the heliocentric idea. What is clear is that Copernicus was fully aware that in suggesting such a possibility he was challenging the authority of Aristotle and therefore the dominant Catholic Church establishment. He was therefore careful with whom he shared his ideas. At the age of 34, Copernicus did publish a handwritten pamphlet outlining his heliocentric model. He distributed it to friends in the scientific community but otherwise he kept his work hidden. In fact, he died in 1543 without ever seeing a final published copy of his great work *De Revolutionibus Orbium Coelestium*, in which his new theory was fully described.

If any justification of Copernicus' extreme caution is required, one only has to consider the tragic case of the writer Giordano Bruno. Bruno was infected with the spirit of Copernicus' work and went so far as to suggest that perhaps the stellar sphere itself was only a creation of man's imagination. In 1600 Bruno was burned at the stake as a heretic for his blasphemous suggestion. To head off negative reactions, the publisher of *De Revolutionibus Orbium Coelestium* added a book preface of which Copernicus was apparently unaware. The publisher wrote that the concept of placing the Sun at the center of the cosmos should be viewed as a calculational tool for convenience, not as a statement regarding reality. However, it is important to note that Copernicus himself was quite clear that placing the Sun at the center was a statement about "truth."

Copernicus' heliocentric theory was remarkably simple compared to the complexity of the Ptolemaic system. It had the additional benefit that it offered a particularly simple explanation for the occurrence of retrograde motion. In Copernicus' model the Earth revolves around the stationary Sun once per year. The other planets orbit around the Sun but with different size orbits and at different rates—Mercury and Venus orbit inside Earth's orbit and at a faster rate whereas Mars, Jupiter and Saturn orbit beyond Earth's orbit and at a slower rate. In fact, it was with the work of Copernicus that the true order of the planets was first correctly identified.

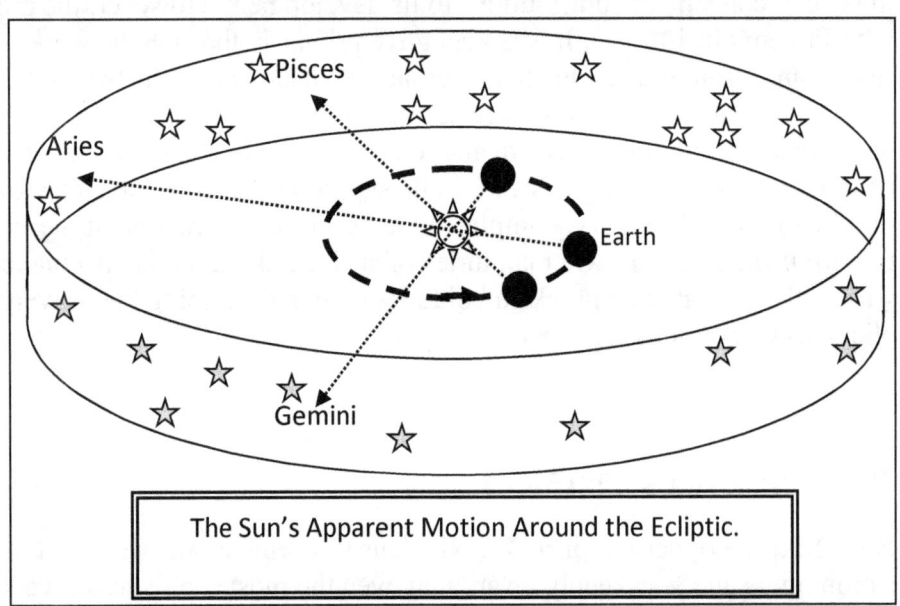

The Sun's Apparent Motion Around the Ecliptic.

At the same time it revolves around the Sun, the Earth spins on its axis once every day. This explains the daily motions of the Sun and stars as resulting not from the motion of the stellar sphere but rather from the motion of the Earth itself. The apparent motion of the Sun along the ecliptic is also naturally accounted for in this system.

As the Earth moves along its annual path around the Sun—in a counterclockwise direction as viewed from the North Star—the Sun appears to shift its position from right to left against the background of fixed stars.

The seasonal variations in the Sun's apparent motion were accounted for by tilting the Earth's spin axis at about 23^0 with respect to its orbital plane. During our summer, the Northern Hemisphere is tilted toward to the Sun, which elevates the Sun's apparent path in the sky. During our winter the

Northern Hemisphere is tilted away from the Sun, which lowers the path of the Sun in the sky and makes the sunlight hit the Earth less directly.

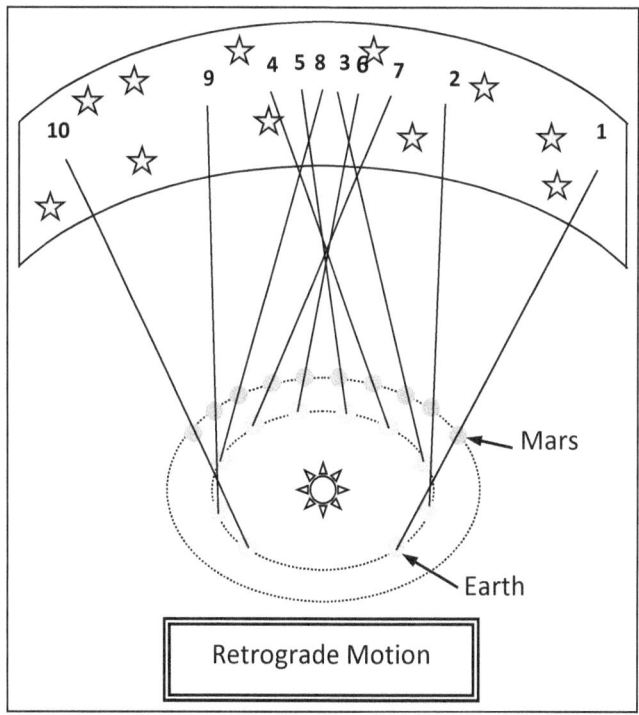

Retrograde Motion

The most remarkable feature of Copernicus' system was its simple explanation for the retrograde motion of the planets. The apparent "backing up" of the planets against the background stars occurs as a result of the differing rates at which the planets orbit the Sun. For instance, Mars takes 1.88 Earth years to orbit the Sun once and therefore gets regularly overtaken by the Earth, which completes the journey in only one year. When the Earth and Mars are on the same side of the Sun and the Earth is overtaking Mars, Mars will appear to back up relative to the stars. This explanation also accounts for the observed maximum brightness of Mars during its retrograde motion—it is this point in the orbit that Mars is closest to Earth.

Although most historians remember the revolutionary character of Copernicus' ideas, careful consideration of his work also reveals the conservative nature of his character. Even though he was willing to demote Earth's status from its privileged position at the center of the cosmos, he nonetheless remained committed to the notion of perfectly circular orbits centered on the Sun. Unfortunately, he insisted (for philosophical reasons) upon circular orbits, thus preventing this new model from closely matching observed behavior. It is true that his heliocentric theory provided simpler explanations of many observed phenomena, but the detailed predictions were quite inaccurate. In an effort to improve the predictive power of his model, Copernicus chose to use planetary epicycles with circular planetary orbits around the Sun. Copernicus was then able to generate acceptable agreement between observation and prediction with his model, but at what expense? In the end, Copernicus' model contained more epicycles than Ptolemy's and *was no better at accounting for observations* (i.e., it was no more accurate). However, despite its lack of predictive power, the new heliocentric model represented a certain aesthetic elegance that many interpreted as a move towards "truth." Even in modern times, the tacit faith in simplicity of scientific explanation has often served as a principle in evaluating new scientific theories. This principle is sometimes referred to as *Ockham's razor* after the 14th century English Franciscan who espoused the notion that the search for truth is the search for simplicity. In other words, given two competing theories, the simplest theory is probably closer to the "truth."

Copernicus recognized that there would be a great deal of opposition to his ideas from both religious and scientific circles. There was little he could do to answer those who objected to his ideas on religious grounds. But there was one serious criticism from scientists. If the Earth were indeed in motion about the Sun, why then had no one ever successfully observed stellar parallax, which would necessarily result from the Earth changing its position while orbiting the Sun. To answer this challenge, Copernicus boldly—and correctly—proposed that the stars are simply too far away to observe any parallax shifts.

It was at about this time that the fundamental tenets of Greek astronomy began to unravel. In 1572 a new star appeared in the sky, known today as Tycho's supernova. Astronomers of the day might have easily dismissed such a thing as an atmospheric event except that careful observations, taken from different points on the Earth, showed that this new star was definitely located beyond the orbit of the Moon. Then, in 1577, a comet suddenly appeared and was similarly demonstrated to be beyond the orbit of the Moon. Both of these observations seriously called into question both the immutability of the heavens and the very existence of crystalline celestial spheres. It would have been hard to ignore the degree to which these observations undermined the philosophical foundations of geocentric astronomy which, despite the revolutionary work of Copernicus, was still very much intact.

> **Box 10-1: Checks and Balances**
>
> Science is sometimes thought of as being composed of two parts: observation and theory. In astronomy, both aspects are of equal importance for making scientific progress. In one sense, experimental physicists can be a real pain in the side of theorists inasmuch as one careful observation can ruin an otherwise wonderful theory but, in a sense, it is the experimentalist's job to find the shortcomings of theories so that better ones may be found. In fact, it has been said that the farther a measurement is from theory, the closer it is to a Nobel Prize. This may be a slight exaggeration, but it is indeed true that often the most important measurements are the ones that, with a high degree of reliability, rule out the current theory and force us to look for new and better explanations. Theoretical astronomers could never have made the leap from the geocentric to the heliocentric theory without volumes of accurate observational data.

10.2 Tycho Brahe

The most import technical observations of the Renaissance came from a gifted and diligent observational astronomer by the name of Tycho Brahe. As well as demonstrating the surprising distances of a supernova and a comet as described in the previous section, he produced what would become the definitive observational catalog of nightly planetary positions on which theorists could base their work.

> **Box 10-2: Tycho the Eccentric**
>
> Any discussion of Tycho Brahe would certainly be incomplete without commenting on him as a person as there are a number of amusing stories surrounding his life. One story concerns a sword duel in which he fought a fellow scholar over some insult (one account has it that it involved a mathematical proof, another involves a lady). In the fight, Tycho's nose was permanently disfigured and, for the remainder of his life, he wore false noses made of gold and silver. Another legend relates that Tycho died under interesting circumstances in 1601. At a nobleman's home for dinner, Tycho ate and drank more than his fill and, as it would have been rude to leave before his host, he refused to leave the table to make use of the facilities. Several days later he died of peritonitis caused by internal injuries suffered that evening. Finally, it is interesting to note on a rather famous portrait of Tycho, one showing the crests of the various branches of his family, the names Rosenkranz and Guildenstern are prominently featured. Is it possible that a British playwright was familiar with this portrait and selected these for the names of characters in his play *Hamlet*?

Tycho served as the court astrologer-astronomer (now a contradiction in terms) to the King of Denmark, who, in 1576, gave him the island of Hven and sufficient funds to build the world's most elaborate observatory. This was prior to the invention of the astronomical telescope. Using precise instrumentation and careful naked-eye viewing, Tycho was able to measure the exact positions of astronomical objects to an accuracy that corresponds roughly to the thickness of a dime held at arm's length—about 2.5 times more accurate than any previous measurements. At least as important as the great accuracy of the measurements was the fact that he maintained a rigorous program of nightly observation for more than 20 years and thus gathered a tremendous store of astronomical data. In interpreting his data Tycho used a curious model in which the Earth was stationary but all of the other planets revolved around the Sun, which in turn revolved around the Earth. This model retained the aesthetics of the Copernican system without the need of accepting a moving Earth.

Another of Tycho's influential contributions to astronomy was to recognize the talent of a young mathematician, astrologer, and mystic named Johannes Kepler. It was on Tycho's recommendation that in 1601, following Tycho's untimely death, Kepler was appointed as imperial mathematician to the Holy Roman Emperor. Tycho's volumes of observational data served as the foundation for Kepler to finally uncover the correct description of the planetary orbits around the Sun.

10.3 Johannes Kepler

Today the name Johannes Kepler (1571-1630) is still associated with his three laws of planetary motion. To a large extent, these laws accurately describe the true motions of the planets—not perfect circles as imagined by Copernicus but another mathematical shape called an ellipse. However, before describing Kepler's laws of planetary motion, we consider a question that has remained unanswered until now: How does one determine the true orbits of other planets by observing planetary positions in the sky?

Kepler immediately abandoned the geocentric model. He was sufficiently persuaded by the simplicity of the Copernican model to accept it as a starting point. He further assumed that the Earth's orbit is nearly a perfect circle, precisely centered on the Sun. This was a very bold (and possibly dangerous) assumption, which turns out to be nearly correct. This was very important because it meant that Kepler could carry out some simple graphical calculations to determine the shape of true planetary orbits.

Box 10-3: Locating a Forest Fire

Kepler's method for determining the position of Mars on a single date is similar to a method that might be used to determine the location of a forest fire. Imagine two groups of firefighters separated by about 5 miles who each spot the same forest fire at the same time and radio back to base to report it. Both groups know in what direction the fire is from their site but they do no know the distance. However, someone at the base can plot the position of the first group on a map and draw a line in the direction that the group sees the fire—the source of the fire must lie somewhere along that line. The same thing is done for the second group's position and direction to the fire, which produces a second line that will intersect the first. Where the two lines intersect MUST be the location of the fire!

As illustrated in the figure on the opposite page, Kepler's approach was very clever. He knew that the Earth goes around the Sun once every 365.25 days, and was able to decipher the correct orbital periods for the other planets as well. For instance, Mars goes once around the Sun every 687 days—about 1.88 Earth years. Suppose that Mars were observed on January 1, 1570. By observing only its position in the sky it would be impossible to know how far away it was. If Mars were then observed again 687 days later, in mid-November of 1571, it would be in exactly the same place as it was on January 1, 1570 having completed one trip around the Sun. However, the apparent direction to Mars would be different because the Earth would be in a different position than for the first observation. With the simplifying assumption that the Earth's orbit is circular, the intersection of the two lines of sight could be used to accurately determine Mars' true position on January 1, 1570.

Obviously this triangulation method gives only one point on Mars' orbit whereas determining a planet's orbit requires knowledge of its position throughout a complete cycle. The beauty of Tycho's impressively modern scientific approach to data collection is that he eventually provided Kepler with hundreds of precise observations of Mars taken over many years. By combining observations in pairs separated by intervals of 687 days, Kepler was able to map out, for the first time, the true shape of Mars' orbit. He found that Mars' orbit is not as circular as Earth's orbit, and that the Sun is not at the orbit's center. On average, Mars is about 1.5 times farther from the Sun than the Earth. Having determined the shapes of the planetary orbits graphically, Kepler's next challenge was to devise a mathematical description of the orbits—it was not good enough merely to call them squashed circles.

Box 10-3: Kepler the Mystic

During the Renaissance scientists devoted considerable effort to avenues of research and inquiry that might seem, by today's standards, to have been quite unscientific. A good example of one of Kepler's less "scientific" pursuits was his search for a reason for the number of planets in the Solar System (then thought to be only six) and their relative spacings. In an argument similar to one made by the Pythagoreans some 2000 years earlier, Kepler imagined a deep connection between the five perfect solids (regular solids, such as the tetrahedron or the cube, that can be constructed from regular polygons) and the spaces separating the six planets. In fact, he constructed an elaborate model in which each planet was enclosed in a perfect solid, which had to fit perfectly into the larger solid for the next planet. By adjusting the sizes and packing order of the solids, he was able to produce reasonable agreement with Copernicus' relative orbital sizes. Kepler was also a strong believer in the "music of the spheres" in which the motions of the planets and stars generate a type of music. He even wrote out the notes played by each of the planets. This sort of work would hardly be considered science by modern standards, but it does highlight just how different the intellectual environment was during the 1600's.

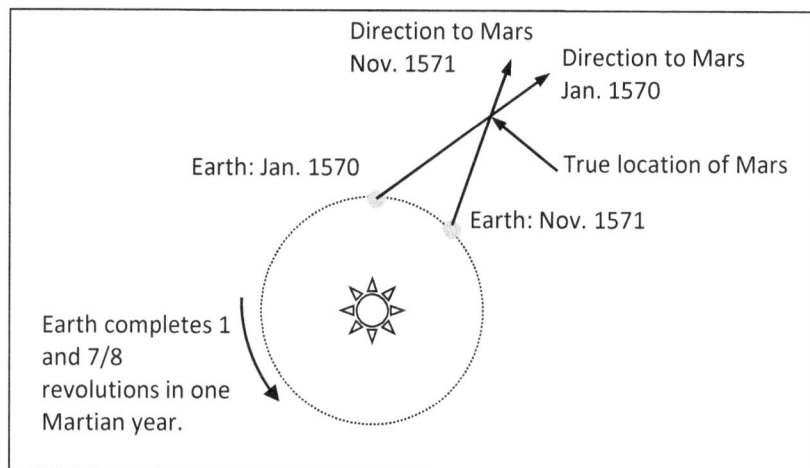

The story of Kepler's progress towards a final solution to this puzzle is one of false starts, endless calculations (no computer!), and eventually triumph. His results were indeed non-intuitive and surprising. The incredible realizations at which Kepler arrived after years of study are summarized in Kepler's three laws of planetary motion.

First Law: All planets follow an orbit around the Sun that is mathematically defined as an ellipse with the Sun at one focus of the ellipse; the other focus is empty.

An ellipse is simply a stretched circle that is longer in one direction than the other. Each ellipse has two focus points that lie along the long axis of the ellipse. As a planet moves along its elliptical orbit, its distance from the Sun is constantly changing. Although the orbits are often drawn as being very "elongated," most of the planets' orbits are *very close* to circular. The amount that each orbit has been elongated from a perfect circle is called the ellipse's **eccentricity**. The Earth's orbit is nearly a perfect circle so it has a very small eccentricity (e=0.017) whereas Mars' orbit is much more eccentric (e=0.094).

Kepler's Second Law: The speed at which each planet orbits the Sun varies so that a line drawn from the planet to the Sun always sweeps out the same area in the same time.

This is perhaps the least intuitive of Kepler's laws. As shown in the figure below, each planet speeds up when it is closer to the Sun so that its swept out area per uniform time interval remains constant throughout the orbit. Likewise, when this same planet is far from the Sun in its orbit, it moves much slower than when it was close to the Sun.

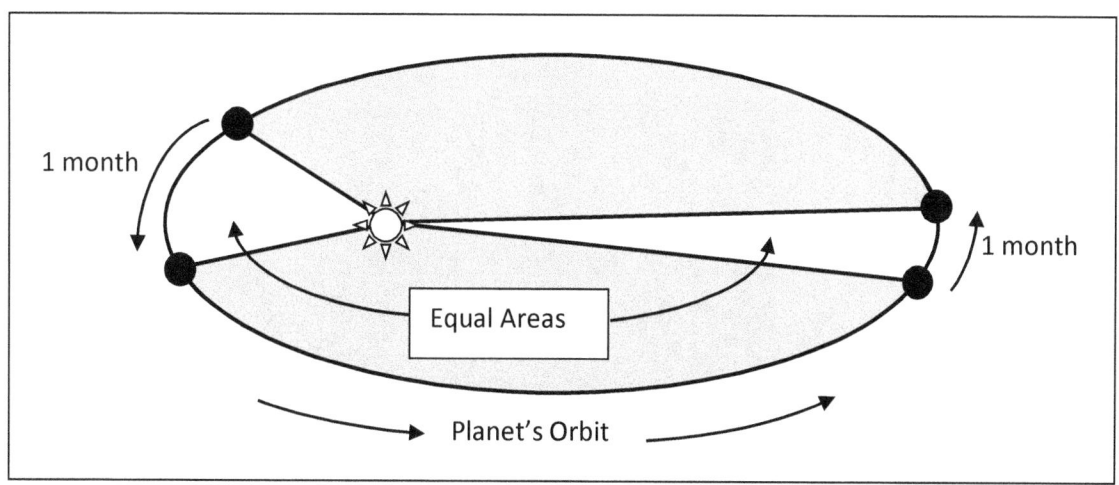

Kepler's Third Law: If you take the orbital period (called T) of a planet and square it (T^2 or TxT) you will get the same value as if you take the planet's average orbital radius (r) and cube it (r^3 or r x r x r) if the period is expressed in Earth years and the orbital radius is expressed in terms of the Earth's orbital radius. ($T^2=r^3$)

This law dictates that planets that have large orbits around the Sun will take longer to orbit than planets that orbit near the Sun. This law has the additional benefit of providing an easy equation to use *($T^2=r^3$)*. The Earth provides a trivial example with $1^2 = 1$ and $1^3 = 1$. For Mars, T=1.88 and r=1.5 giving T^2= 3.5 and r^3= 3.5 in agreement with the Kepler's third law. That the law works for the Earth is no surprise but the fact that it works for Mars, with the measurements in terms of the Earth's properties, demonstrates that this is the only of the laws that actually connects the motions of the different planets. If one measures the period of any object orbiting the Sun—including asteroids and comets—then Kepler's third law allows the prediction of orbital size. It may well have been this connection between the motions of the planets that suggested to Newton that something even more fundamental must underlie and control the motions of all planets—something known today as Newton's law of gravity.

Chapter 11
Founders of a Science

Many scholars maintain that the modern practice of physics was seeded by Galileo Galilei and grew into all its glory under the guidance of Sir Isaac Newton. In fact, many university physics courses do little more than study the ideas of these two great scientists. However, Galileo's contributions to astronomy remain distinct from other important work he did in the study of basic mechanics (the study of how and why things move). It is in the writings of Isaac Newton that we see perhaps the greatest intellectual revolution in his joining together the fields of the terrestrial (Earthly) and celestial (heavenly) realms in a single unifying framework. This revolution eliminated once and for all the artificial distinction between Earth and heavens and showed that both operate under the same set of basic scientific laws.

11.1 Galileo (1564-1642)

Galileo Galilei was a contemporary of Johannes Kepler, but his contributions to astronomy were quite different. Galileo focused on the observations of stars and the characteristics of the planets that he was able to see with his telescope. Whereas the work of Copernicus and Kepler dealt exclusively with understanding the motions of the objects in the heavens—called **celestial mechanics**—Galileo's work in astronomy focused instead on describing the physical characteristics of those objects, which required the construction of an astronomical telescope. While Copernicus and Kepler were finally displacing the Greek models of heavenly motion, Galileo's discoveries were attacking the philosophical underpinnings of the Greek astronomers as well as providing additional evidence in support of the Copernican view. Galileo made at least seven monumental observations, each of which had specific consequences in terms of creating a new cosmology finally freed from the orthodox Greek philosophy.

> **Box 11-1: Italian Galileo Galilei**
>
> Many people believe that Galileo invented the telescope. Actually, a lens maker in Holland is credited with the invention in 1608. Galileo did, however, learn of its invention and quickly fashioned his own. He was the first person to use a telescope to systematically study the heavens and, more importantly, he had the intellect to appreciate the importance of what he was seeing and to write prolifically about it. His writings are found in a widely accessible book called *Siderius Nuncious*, or *The Starry Messenger*.
>
> Also, Galileo was not condemned of heresy by the Inquisition for believing that the Earth goes around the Sun. Galileo did indeed have ongoing difficulties with the Catholic Church, which began with the publication of *Siderius Nuncious*. In 1616, Pope Paul V ordered Galileo to cease his outspoken (and often tactless) defense of the Copernican system. Later, Galileo attracted the attention of the Inquisition by his apparent ridicule of Pope Urban VIII in his book *Dialogo Dei Due Massimi Sistemi* (*Dialogue Concerning the Two Chief World Systems*). Galileo presented his scientific arguments in the form of a play—dialogues between proponents of the geocentric and heliocentric viewpoints. Galileo represented the geocentric views—those that were espoused by the Pope—through the character Simplicio, who was clearly meant to portray the fool. Galileo was ultimately condemned for disobeying the orders of 1616, a charge that might be considered a technicality.
> In 1992, Galileo was expiated (religious reinstatement) by Pope John Paul II (in no small part) to formally recognize that religious faith can adapt to a changing knowledge base. In the formal expiation proceedings Pope John Paul II recounted the 300 year old words of Galileo, and the 2000 year old words of Aristotle: when the interpretation of one's guiding scriptures are no longer representative of known truths, then it is time to reinterpret one's guiding scriptures.

Features on the Moon

The religious thinking of the day assumed that the Moon was rather like a perfect jewel in the sky as befits its heavenly location. This was in sharp contrast to the Earth, which was the realm of disease, evil, and human frailty. Galileo's observation that the Moon is covered with jagged mountains and sinuous valleys with a rugged and cratered surface certainly removed it from its lofty pedestal of perfection.

The Milky Way

To the naked eye, the bright band that stretches across the sky called the Milky Way is nothing but a smeared out band of light. However, using his telescope, Galileo was able to see that the Milky Way is composed of almost countless dim stars. This added considerably to the some 6000 known stars in the Universe and thus further challenged the Earth's privileged position at the center of the Universe.

Disk-shaped planets

One of the first things that the owner of a new telescope quickly learns is that all stars, no matter how much you magnify them, still look like points of light. Planets, on the other hand, have a definite size when viewed through a telescope—they appear disk-like. Galileo observed this difference, thus providing evidence that is was more than just their *motion* that made stars and planets different.

Saturn is not round

Galileo's primitive telescope was not able to fully resolve Saturn's rings but he was able to distinguish a bulge around its center, which showed Saturn not to be round—Galileo described Saturn as bean-like. This was another challenge to the Greek ideal of the heavenly perfection of circles and spheres. It is interesting that Galileo later began to doubt his own observational abilities because he was no longer able to see the bulge of Saturn—of course he had no pictures of his first observations to verify their validity. We now understand that, during his later observations, he happened to be viewing Saturn's rings directly from the side (like looking at the thin edge of a dime) making them seem to disappear.

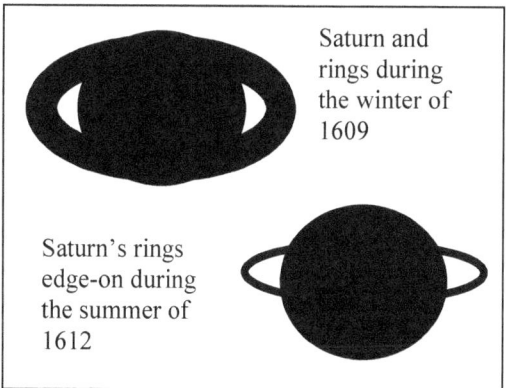

Saturn and rings during the winter of 1609

Saturn's rings edge-on during the summer of 1612

Sunspots

Turning his telescope on the Sun, Galileo observed dark spots that appeared to be permanent markings on its surface. This not only provided additional evidence that celestial objects are not unblemished spheres but, by following the sunspots' motions across the Sun's surface, he was able to infer that the Sun rotates about its own axis in about 26 days. Besides reporting the existence of blemishes on the otherwise perfectly smooth Sun, this was important because one of the arguments against the Copernican system was that a body as large as the Earth would tear itself apart if it were to spin. If the Sun could spin about its axis without tearing apart so could the Earth.

Image Courtesy of NASA and Yohkoh

★ ★ ◯ ★ ★	Monday
★ ★ ◯ ★ ★	Tuesday
★★ *◯ ★	Wednesday
★* ★ ◯ ★	Thursday
★★ ◯ ★	Friday

The Moons of Jupiter

When Galileo pointed his telescope at Jupiter, he discovered four new stars very close to the planet. Following these new "stars" night after night, it soon became clear that they were not stars at all as evidenced by their changing positions. Galileo surmised that these "stars" were in fact moons in orbit about the planet Jupiter. These four moons—Io, Europa, Ganymede, and Callisto—are now called the Galilean moons (we now know that Jupiter has at least 18 moons). His discovery of Jupiter's moons was particularly important evidence in support of the Copernican system for two important reasons. First, this observation demonstrated conclusively that there could be more than one center of revolution in the Solar System. Those still clinging to a geocentric model believed that, like the Earth's Moon, all objects moved around the Earth. Galileo clearly demonstrated that objects could orbit other objects besides Earth.

Second, the discovery of Jupiter's moons provided evidence to counter one of the other popular arguments of the geocentrists. They were suggesting that the Copernican system was untenable because our own Moon would never be able to keep up to a moving Earth. That Galileo's four Jovian moons could apparently keep up with Jupiter—a planet that everybody agreed was in motion—countered this popular objection to Copernicus' heliocentric model.

> **Box 11-2: More than 2000 years too late?**
>
> Galileo is generally credited with discovering Jupiter's moons, and yet their recorded observation may actually predate Galileo by some 2000 years. The Chinese astronomer Kan Te apparently noted that Jupiter was followed by a small reddish star attached to it. Under ideal conditions, the brightest of Jupiter's moons can be visible to the naked eye, which may mean that Galileo's observations of Jupiter's moons were actually the second on record, coming two millennia after the first.

The phases of Venus

Using his telescope, Galileo observed that Venus clearly displays phases like Earth's Moon, ranging from a thin crescent to a fully lit disk. If Venus were always located closer to Earth than the Sun, as required of the Ptolemaic model, then Venus would always be illuminated from the back, allowing us to observe only various degrees of a crescent phase. Galileo clearly saw Venus in nearly full phases that can only occur if Venus is sometimes on the opposite side of the Sun from us, so Venus was clearly orbiting the Sun. If Venus could do so, what about the other planets? Put bluntly, Ptolemy was wrong and Copernicus was right!

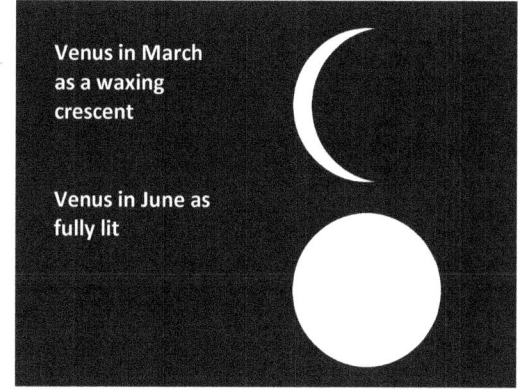

Galileo, imprisoned in his villa under house arrest for the last ten years of his life, died on January 8, 1642. Eleven months later, on Christmas day of that year, arguably the single most influential physicist of all time was born. His name was Isaac Newton.

11.2 Isaac Newton (1642-1727)

It is almost impossible to overstate the importance of the contributions to contemporary science made by Isaac Newton, which were in no way confined to astronomy. Newton's work in optics forms the foundation for modern theories of light and, in trying to understand the details of gravitational attraction, he was momentarily diverted by the need to invent an entirely new type of mathematics—the calculus. Although the details of Newton's work are beyond the scope of this course, it is important to understand where his work fits into the history that we have been tracing.

Newton was interested in the motion of the Moon, which he envisaged to be in a constant state of free-fall around the Earth. This can be understood by imagining standing atop a tall mountain and firing ever more powerful canons towards the horizon (see figure). The faster the shell is fired, the farther it lands from the base of mountain. But, and this requires some real imagination, if the ball is fired so fast that the Earth actually curves down significantly before the ball lands, then the ball has even farther to drop and "bends" around the Earth. Even faster balls go yet farther around until a ball is fired so fast that, even though it keeps bending towards the Earth, it can't get any closer because the surface of the Earth keeps bending out of the way. This ball is said to be in orbit around the Earth.

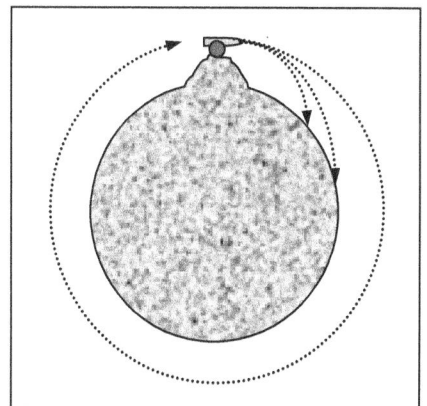

If you are not accustomed to the idea, it is hard to think of something in circular orbit as falling. Yet, this is now considered to be the most correct interpretation. Newton's genius was to "see" the Moon as falling in much the same way as we have described the cannon ball. Because he knew the details of the Moon's orbit around Earth, Newton was able to determine that in one minute the Moon's path bends by only about 17 feet from being a straight line. Newton understood that this also meant that any object, released from rest at the 400,000 km orbiting distance of the Moon, would fall only 17 feet towards the Earth in one minute.

> **Box 11-3: The Space Shuttle**
>
> Compared to the Moon, which orbits at a distance of about 60 times the Earth's radius, the Space Shuttle orbits very close to Earth. In fact, its orbit is only about an additional *one tenth* of the Earth's radius above the Earth's surface! The Space Shuttle travels around the Earth about once every 90 minutes, traveling at about 18,600 miles per hour. In one minute, the shuttle falls about 11 miles. This is the amount by which it would get closer to the Earth if it were falling straight down. In that same one minute, the shuttle travels a distance of about 310 miles along its path. If the shuttle actually went in a straight line (tangent to its path) for a distance of 310 miles then the surface of the Earth, because it is bending away, would actually get 11 miles farther away. In "falling" 11 miles closer to Earth the shuttle is actually just keeping pace with the curve of Earth's surface so that it maintains a constant distance above the ground.

At this stage, Newton made the bold assumption that the same force that is responsible for falling apples is also responsible for the "falling" Moon. The difference between how far an apple falls in one minute (about 11 miles without air resistance!), and how far the Moon falls (about 17 feet) is a result of the weakening of the Earth's gravitational force with distance. Newton showed that the 17-foot drop of the Moon is the result of a force that is about 1/3600 as strong as that experienced by an apple (if you correct for the different masses of an apple and the Moon). As the Moon is about 60 times farther from the center of the Earth than an apple on the Earth's surface, Newton concluded that the force, called gravity, must have the property that every time the distance between the objects is doubled, the force gets 4 times weaker. We call this a "one over R squared" force, where R is understood to be the distance between the centers of the interacting bodies. This law is referred to as Newton's law of universal gravitation and its discovery marked the beginning of a whole new way of conceptualizing the universe. Newton revealed that there is nothing special about the heavens—the future develops there according to exactly the same rules as here on Earth. For the first time, we see the unification of **terrestrial** and **celestial mechanics** into a single framework.

Newton then went further and wrote that the same laws used to describe both apples and the Moon should also govern the orbits of all the planets around the Sun. Whereas Kepler had only described the motions of the planets, Newton explained the reason for those motions. He was able to start from only a few basic laws and show that all of the properties that Kepler had described were a necessary consequence of the underlying physical principles. If Kepler's laws had still been unknown, Newton could well have deduced them directly from his theory and then checked the results against the observational data.

Finally, with the aid of Newton's insights, we can address an issue that is left somewhat vague in our discussion of the masses of stars: How can we measure the mass of a planet or a star? One property of orbital motion that Newton demonstrated is that the rate at which a small object orbits a much larger one—a moon around a planet, a planet around a star, or even a small star around a larger companion star—depends only on the mass of the larger object and the radius of the orbit. Most important, it does not depend on the mass of the orbiting object. To make sense of this you might recall seeing astronauts during a space walk. Once the astronaut lets go of the Space Shuttle, she/he becomes a very small object in orbit around the Earth whereas the Shuttle itself is a relatively large object in the same orbit. If these large and small objects did not orbit at the same rate then they would quickly become separated, which they do not. Either a space ship or a comet would take the same amount of time to orbit the Earth if it were located at the orbital distance of the Moon.

> **Box 11-4: Laws of Orbital Motion**
> 1. *The length of time it takes for a body to go around once (the orbital period) does not depend on the mass of the orbiting object.*
> 2. The orbital period of an object depends only on the radius of the orbit and the mass of the central body around which it is orbiting.

The mass of the Sun can be deduced by knowing the radius and period of any planet's orbit around the Sun. We weigh the Earth by knowing the period and radius of our own Moon's orbit. Jupiter is weighed by noting the period and orbital radius of any of its moons. In fact, any time a smaller object in space has its path altered by the gravitational force of a larger object, the mass of the larger central object can, in principle, be determined by simply measuring the orbital period and the orbital distance of the smaller body orbiting it.

Chapter 12
The Search for the Outer Planets

For several millennia, the focus of celestial mechanics—the study of the motions of the heavens—was on finding a correct *description* for the motions of the stars and planets. Once Kepler uncovered a coherent and consistent description, summarized in his three laws of planetary motion, then scientists turned their attention to seeking an underlying cause. Given the natural order seen in Kepler's laws, it was natural to look for a common cause for planetary motions at a more fundamental level. Isaac Newton, by combining his law of universal gravitation with the laws of basic physics, was able to show that Kepler's laws were a necessary consequence of these more fundamental laws concerning the workings of nature. The powerful laws of Newton finally explained why a planet located where Mars is moves the way it does. What Newton's laws could not explain was *why* Mars was located exactly where it was and not somewhere else.

12.1   Titius-Bode Rule for Planetary Distances

Kepler attempted to explain the arrangement of planetary orbits by imagining each of the planets to be separated from the next by one of the five perfect solids. Another theory on the spacing of the planets was first proposed in 1766 by a German mathematician and astronomer named Johann D. Titius. This work lived in relative obscurity until Johann Bode from the Berlin Observatory popularized it in 1776. The Titius-Bode Rule is remarkably simple and accurate. Start with the series of numbers 0,3,6,12,24,48... where each number after 3 is simply twice the previous number. To each of these numbers, add 4 and divide by 10 yielding the series 0.4, 0.7, 1.0, 1.6, 2.8, 5.2, 10.0, 19.6, 38.8, etc. These numbers correspond remarkably well to the average radii of the planetary orbits expressed in AU.

START	0	3	6	12	24	48	96	192
ADD 4	4	7	10	16	28	52	10	196
DIVIDE BY 10	0.4	0.7	1.0	1.6	2.8	5.2	10.0	19.6
Corresponding Planet	Mercury	Venus	Earth	Mars	Asteroid Belt	Jupiter	Saturn	Uranus

For instance, Mercury actually orbits at 0.38 AU, Venus at 0.72 AU, Mars at 1.5 AU, etc. The only real discrepancy is that Bode's Rule predicts the existence of a planetary orbit with a radius of 2.8 AU, which lies between the orbits of Mars and Jupiter. However, it turns out that in 1801, when looking for the missing planet predicted by Bode's Rule to be at 2.8 AU, Giuseppe Piazzi found the first member of the large asteroid belt that lies precisely in this region. Piazzi named this first asteroid Ceres. Ceres is large compared with the other tens of thousands of asteroids that reside in this asteroid belt, but it is much smaller than any planet. Perhaps more interesting though is that Bode's Rule also predicts a planet with an orbital radius of 19.6 AU to reside beyond what was then thought to be the outermost planet—Saturn.

12.2 William Herschel's Discovery of Uranus

In the mid 1700s a young musician began his scientific career when he discovered how relaxing it could be to sit in the evening and work out solutions to calculus problems—many of you have no doubt found this to be quite relaxing as well. William Herschel then turned his attention to astronomy and soon became one of the most well equipped amateur astronomers of his day. In 1781, stellar parallax, though a necessary consequence of a heliocentric solar system, had not yet been observed. Herschel's primary goal was to make careful observations of the apparent positions of stars to verify the existence of stellar parallax. His painstaking observational program, carried out over an extended period of time did pay off (but not in the way he had imagined). On March 13, 1781 he noted:

> *In examining the small stars in the neighbourhood of H Geminorum I perceived one [a star] that appeared visibly larger than the rest; being struck with its uncommon appearance I compared it to H Geminorum and the small star in the quartile between Auriga and Gemini, and finding it so much larger than either of them, suspected it of being a comet.*

Herschel originally suspected that he had discovered a comet and quickly reported his findings to the astronomical community. Before long, the most powerful telescopes in Europe were trained on this new object in the heavens. However, they soon realized that Herschel's object, though showing a visible disk (not point-like like a star), showed no tail and lacked the characteristic "fuzzy" edges of a comet. As astronomers began to track the motion of the object across the sky it became clear that it was moving too slowly to be a comet. Applying Newton's laws of gravity, astronomers were quickly able to demonstrate that this new object had all of the characteristics of a new planet, which Bode named Uranus. Herschel actually wished to name the planet *Georgium Sidus*, or George's Star, in honor of King George III of England. In the end, the new planet retained its current name of Uranus.

Within about two years, enough data had been gathered to establish the basic parameters of the new planet's orbit. It was found to reside about twice as far away from the Sun as Saturn at an average radius of 19.6 AU, in amazing agreement with the prediction of Bode's Rule.

Early success in understanding Uranus' orbit was to be short-lived. By 1830 there was a general consensus among astronomers that there were indeed small, but measurable, discrepancies between the observed orbit of Uranus and the predicted motion based on Newton's laws. There seemed to be four possibilities:

Box 12-1: Is It a Matter of Luck?

It is interesting to note that Herschel's observation of Uranus has often been attributed to luck, a charge against which Herschel vigorously defended himself. He was correct in pointing out that it was the systematic nature of his observational program that allowed him to "discover" this new planet and just because it was not what he was seeking should not be counted against him. Many important discoveries in science appear serendipitous. Ultimately, however, most discoveries rely on investigators with the necessary knowledge and experience to recognize a new phenomenon when they observe it. In fact, the ability of theoreticians to nail down the orbit of Uranus depended on earlier observations of the planet that were recorded without appreciating the nature of the object. It is now believed that, dating back to 1690, Uranus had been observed and its position recorded as a simple star at least 22 times prior to Herschel's discovery.

- There were mistakes in the data.

- There were mathematical errors in the application of Newton's laws. The calculations were difficult because the effects of the pull from the other planets as well as the major pull from the Sun had to be included.

- Newton's laws were wrong or incomplete.

- There was an additional unseen planet beyond Uranus whose gravitational attraction was perturbing Uranus' orbit.

The majority of astronomers and mathematicians of the day came to favor the hypothesis of an additional planet pulling Uranus out of its prescribed orbit. The difficulty was where to look for this undiscovered planet.

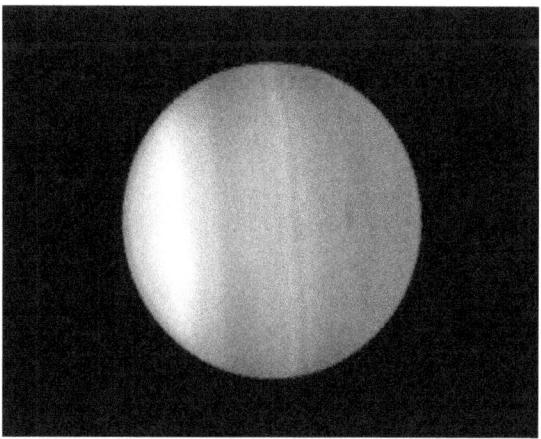

12.3 The Discovery of Neptune by Prediction

The mathematical search for the unknown planet was lead by two scientists working individually, Englishman John Couch Adams and Frenchman Urbain Jean Joseph Le Verrier. Using Newton's law of gravitation and records of perturbations in Uranus' orbit, they began to mathematically predict where this mystery planet, which was responsible for pushing and pulling Uranus off its predicted path, would be located. Both were able to arrive at predictions of the orbit of this new planet and, more importantly, were able to suggest to observers where to look to find it—their recommendations were very similar. Unfortunately they both had difficulty in getting anyone to take them seriously. After several years, in 1846, Johann Gottfried Galle at the Berlin observatory went looking in the constellation of Aquarius, where Adams and Le Verrier suggested, and found the next faint planet. Herschel named this planet Neptune. There was much controversy at the time as to who should get credit for the discovery but, in any case, it *initially appeared* to be a real triumph for the predictive power of Newtonian mechanics. However, we now know that the process of working back from the effect to the cause does not yield unique solutions. There can be many different orbits that will produce the same effects on a neighboring planet. Today we know, from working through the calculations of both Adams and Le Verrier with sophisticated computer models, that it was really a remarkable coincidence that their predicted positions coincided with the real position at that time. It has been shown that their predictions would have put Neptune in quite the wrong place 70 years later, making it difficult to find.

> **Box 12-2: The Search for Vulcan**
>
> Aside from the wobbles in Uranus' orbit that seemed to demand the existence of a new planet, there was another nagging problem in celestial mechanics: the problem of the precession of the perihelion of Mercury's orbit, a problem to which Le Verrier would devote 31 years of study. Mercury's orbit is elliptical, or "egg-shaped." With time, the "point-of-the-egg" appears to rotate—it remains in the same plane but the direction changes like the hands on a clock. It takes more than 227,000 years for the orbit to precess around once. To a large extent, the reason for this was understood as resulting from the gravitational interactions between Mercury and the other planets. There remained, however, a discrepancy that could not be accounted for in this way. Ultimately, Le Verrier proposed the existence of an additional planet orbiting inside Mercury. He called this planet Vulcan because it would be unbearably hot as it would be very close to the Sun. His case was strengthened when, in 1859, an amateur astronomer reported seeing a small planet crossing the disk of the Sun, but of course this was never confirmed. In 1918, Albert Einstein announced his theory of general relativity, which claimed that deviations from Newton's law of universal gravitation could be expected in the vicinity of massive bodies like the Sun. Einstein showed that Mercury was the only planet close enough to the Sun to experience these effects and, in one of his theory's great triumphs, he showed that the observed precession of Mercury's orbit could be predicted exactly within the framework of general relativity.

As more data on Neptune's orbit was collected—this included old recorded observations including one by Galileo—it became possible to determine the details of Neptune's orbit. It became increasingly clear that Neptune's orbit did not agree with the theoretical predictions any better than Uranus' orbit had. It should also be noticed that the orbit of Neptune *is not predicted* by Bode's Rule—the first planet not to fit into this scheme.

12.4 Clyde Tombaugh and The Discovery of Pluto

It was not long before discrepancies in the orbit of Neptune began to suggest that another planet might exist even farther from the Sun. Harvard astronomer William Pickering and eccentric businessman Percival Lowell were the main contenders in the hunt for this new planet. Lowell purchased a mechanical calculator to help with the complex calculations involved in predicting the orbit of the unseen planet presumably responsible for Neptune's orbital discrepancies. Even with this computational aid, it took more than 10 years of work to produce a prediction for the position of the mysterious planet X. Pickering used a less intensive but more approximate technique to solve the necessary equations and was able to produce a prediction in 1908, seven years before Lowell. Unfortunately, attempts to find this new planet continued to come up empty and were halted altogether in 1926 when Lowell died. The search at the Lowell Observatory resumed in the late1920s and, after going through more than 90 million star images captured on dozens of photographic plates, Pluto was discovered by Kansas farm-boy Clyde Tombaugh on February 18, 1930.

In searching for a planet or any other non-stellar object, the goal is to identify a point of light that appears to shift its position relative to the background stars. Using pictures of the stars taken about a month apart, astronomers look for objects that have changed position. Today, this search can be accomplished painlessly with a computer but, in the time that the search was on for Pluto, this had to be done manually. This painstaking task was aided with the use a machine called a *blink comparator*. Flipping a mirror between two photographic plates—two star field photographs showing the same piece of sky on different nights—allowed astronomers to search for objects moving against the background stars. Stars would be in the same position on both plates and appear stationary. A planet, in contrast, would appear to *blink* as its position would be different in the two plates.

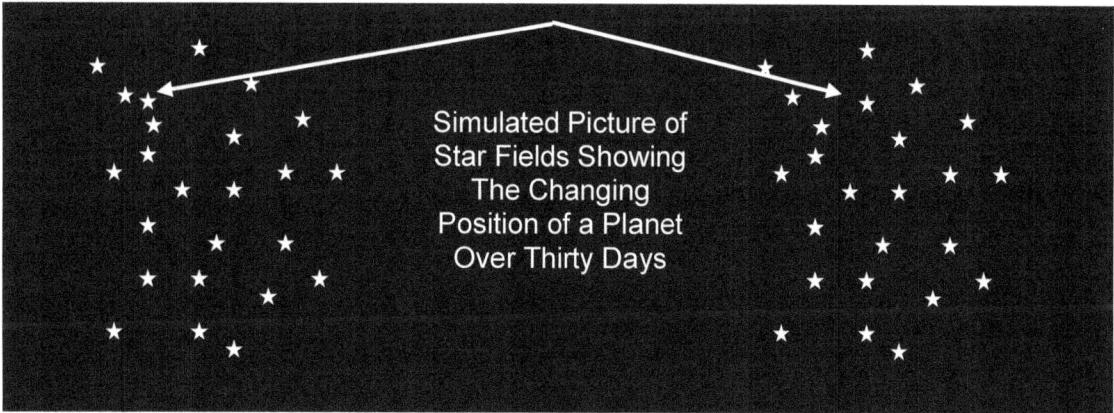

It turns out that Pluto is so very small that it can not possibly influence the orbital path of Neptune. Again, it appears that the discovery of Pluto was pure serendipity and was in no way aided by all of the hard work that was put into predicting its position based on the motion of Neptune. On the other hand, Pluto's position agrees with Bode's Rule!

A recent debate within the astronomical community is whether or not Pluto should even be considered a planet. Its orbit is highly elliptical causing to pass inside Neptune's orbit during part of its journey around the Sun. Also, the plane of its orbit is inclined 17 degrees with respect to that of the Earth, which is much more than any of the other planets. Recently, a new collection of objects (known as the Kuiper belt) has been confirmed to exist beginning at about the orbit of Pluto. Some astronomers consider Pluto to be the largest of the Kuiper belt objects and not worthy of planet status at all. Naturally, Clyde Tombaugh passionately opposed this suggestion until his death in 1997.

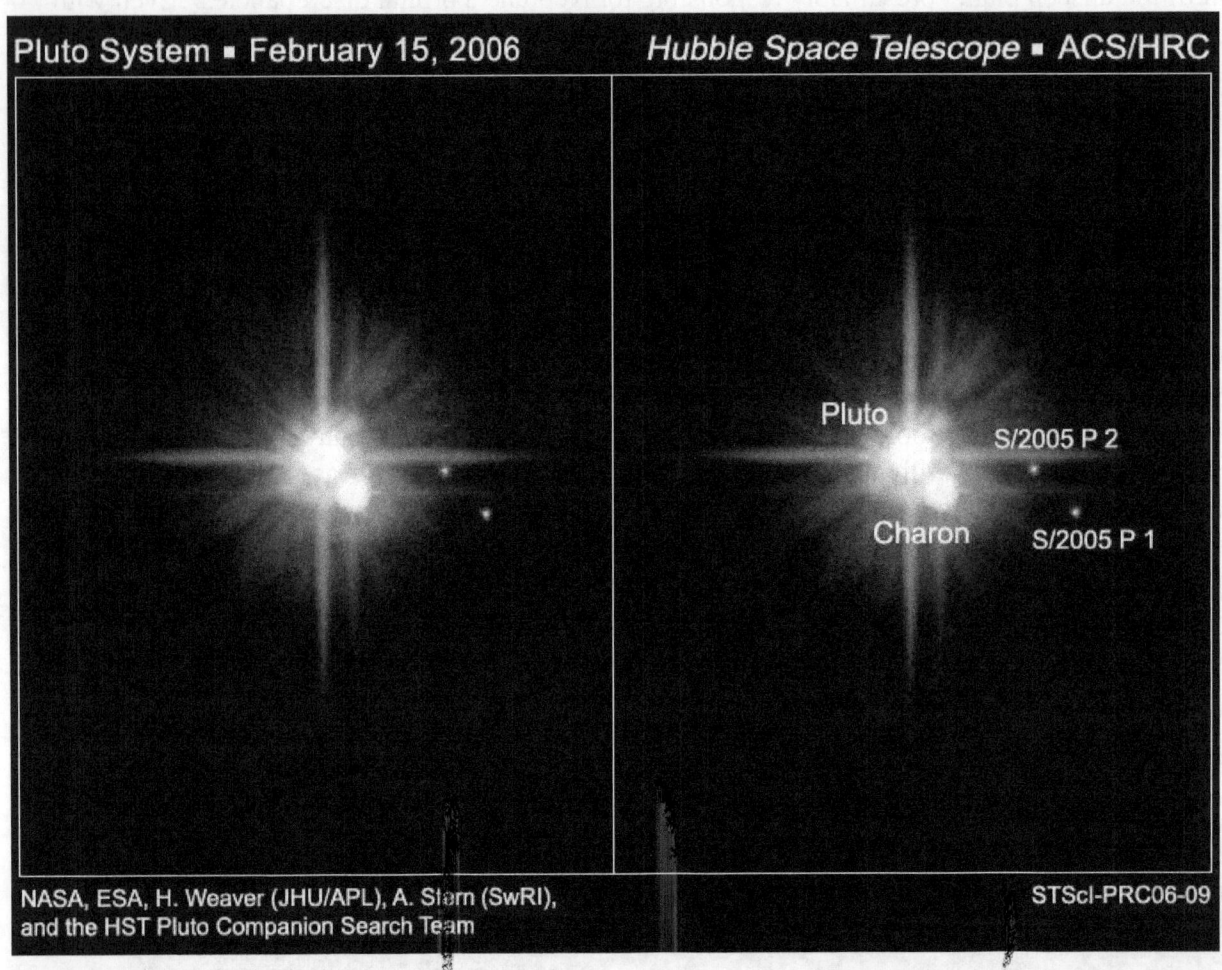

To give you some sense of the enormous distance to Pluto, consider that Pluto's moon Charon, which is nearly a third the size of the planet itself, was not discovered until 1978. And, not until the launch of the Hubble Space Telescope, were images available that clearly showed Pluto and Charon as distinctly different objects. In late 2005 Hubble discovered that Pluto has two additional moons, much smaller than Charon. They are named Nix and Hydra (labeled as S/2005 P 1 and S/2005 P 2 respectively).

Chapter 13
Evolution of the Solar System

Our Solar System is composed of one star, eight planets, over 120 moons, thousands of asteroids, countless comets, dust, and gas. A detailed survey of individual planet characteristics is hardly necessary to appreciate that there are great differences between the properties of the planets with each displaying unique characteristics. Although each planet is interesting in its own way, we will focus on planetary similarities as a guide to understanding the origin and evolution of the Solar System. It is a truism that when scientists are looking to understand "the big picture" they are much more apt to be guided by similarities than by differences.

13.1 Dynamic Properties of the Solar System

To investigate the origin of the Solar System, an inventory of Solar System characteristics is useful. These properties can be divided these into two distinct classes. The first class is the dynamical properties that deal with the motions of the planets and moons—properties that do not depend on the specific structure or composition of the bodies involved. For example, the period of the Earth's Moon does not depend on the Moon's composition; any object located that distance from the Earth would orbit in the same way. These dynamical properties can be observed directly once we have a coherent model for the structure of the Solar System. Listed on the next page are the general dynamical properties of the Solar System, many of which have been known for hundreds of years with the fine details just awaiting better telescopes.

Overall, the Solar System is a flattened disk that revolves about its center in almost perfect circles with its elements spinning in the same counterclockwise sense as the revolution. Its moons orbit the constituent planets generally in the same way. As Newton's laws require none of these properties, we are drawn to seek an explanation or common cause for this consistent regularity.

> ### *List of Dynamical Properties*
>
> - All the planets orbit in nearly the same plane (i.e., the Solar System is flat).
> - All planets orbit in the same direction–counter-clockwise as seen from the North Star.
> - The Sun spins in this same counter-clockwise direction.
> - Most planets and the Sun spin "upright" so that the spin axis is up and down—the Earth is actually tilted about 23 degrees from upright.
> - Most of the planets' moons orbit around the planet equators rather than over the poles.
> - Most of the moons also spin counter-clockwise.
> - The orbits of the planets are nearly circular rather than being highly elliptic. *(Remember that Newton's laws DO NOT favor circular orbits over highly elliptic orbits.)*

13.2 Physical Properties of the Solar System

The physical properties refer to the specific properties of planets and, in general, cannot be observed directly. For instance, scientists agree that the core of our own Earth is made of iron. However, this has never been observed directly—it must be inferred from evidence such as seismic waves spreading from earthquakes. As such, it is worth remembering that our data on the physical properties of planets often represent educated guesses—sometimes very educated and well informed, but guesses nonetheless.

A discussion of the complete physical properties of any one planet could easily fill a textbook, especially if that planet were Earth. We therefore limit ourselves to those properties that are particularly relevant to the consideration of Solar System evolution. The simplest, and most direct, property of a planet to measure is size. Thousands of years ago, Aristarchus determined the relative sizes of the Moon and Sun by first determining their relative distances from Earth. He relied on the basic concept that objects appear smaller if they are farther away. Telescopes can be used to determine the apparent sizes of planets and, since the distances are known, the true sizes can be calculated. The relative sizes of the planets, as well as the sizes of the orbits, naturally divides the planets into two distinct groups of 4 members each.

The four innermost planets (Mercury, Venus, Earth and Mars) are known as the **terrestrial planets**. The rocky terrestrial planets are all about 100 times smaller than the Sun and orbit at between 0.4 and 1.5 AU. The next four planets (Jupiter, Saturn, Uranus, and Neptune) are, by contrast, giant planets. These giant planets average about 10 times smaller than the Sun and orbit between 5 and 30 AU, considerably farther than the terrestrials. Besides their immense size, these planets share many other important physical characteristics and so get grouped together as the Jovian (Jupiter-like) planets. Pluto, with its small size, highly elliptical and tilted orbit, and abnormally large companion moon, Charon, stands out as an oddball amongst the outer planets. On August 24, 2006, the IAU refined the definition of a planet and Pluto is no longer considered a planet.

Based purely on observational data, we can note two additional features that differentiate these two groups of planets:

1. The terrestrial planets have few moons (Mercury and Venus have none, Earth has one and

Mars has two). The Jovian planets, in contrast, have tens of moons each (63 for Jupiter, 34 for Saturn, 21 for Uranus, and 13 for Neptune).

2. No terrestrial planet has rings whereas all of the Jovian planets have rings—the rings of Saturn are the only ones big enough to be seen easily with a small telescope.

Another interesting property to consider is the average densities of the planets. The density can be thought of as how many times heavier an average bucket of material from the planet would be compared to a bucket of water. For example, the density of aluminum is 2.7, silicate rock is about 3.0, iron is 7.8 and lead is 13.4.

Average density, though clearly not a very detailed measure, does offer an additional basis to distinguish between the terrestrial and Jovian planets. The terrestrial planets vary in density from 3.9 to 5.5, suggesting high abundances of iron and rock whereas the Jovian planets fall in the range 0.71 to 1.67 suggesting that they contain mainly lighter elements. In fact, the Jovian planets are often called gas giants, referring both to their large size and gaseous compositions.

Just as spectroscopic evidence was used to identify the chemical composition of distant stars, it can also be used to learn about the much nearer planets. In general, astronomers make the following general conclusions about the physical compositions of the planets:

1. The Jovian planets are composed mostly of hydrogen (75%) and helium (25%) with only traces of anything else. In terms of composition, these planets are very much like our own Sun. Although they consist mainly of very thick, dense clouds of gas (they have no true surface) they might contain small, rocky cores at their centers.

2. The terrestrial planets consist mainly of heavy elements such as we find on Earth: silicon, oxygen, iron, magnesium, etc. They are solid, rocky objects surrounded by just a thin gaseous atmosphere.

Having now completed this brief comparison of the planets, we are ready to begin speculating about how our Solar System came into being. The requirement is that any model we develop must explain the properties that we now observe. One note of caution is that the only data at hand is but a single momentary snapshot in time of the state of our Solar System and this may not be particularly representative. For instance, astronomers now believe that the surface of Mars once flowed with water. Astronomers also think that planetary rings, like the familiar ones around Saturn, are probably temporary features. Furthermore, the composition of Earth's atmosphere is not a result of the physical evolution of our planet, but is rather a byproduct of the life that has formed and thrived here.

13.3 Nebular Hypothesis of Solar System Origin and Evolution

The goal of the previous section was to highlight the incredible degree of regularity found in the Solar System—regularity that is not explained by any laws of physics that we have discussed. Newton was indeed able show that the orbits of the planets should be elliptical, but his laws provide no requirement that they be oriented in any consistent plane. Also, whether we believe it is science or just trickery, we saw with Bode's Rule that the spacing of the orbits does seem to follow a mathematically simple scheme. When the similarities cannot be explained by a set of current constraints, the next place we tend to look is for a common origin. Until relatively recently, there were three competing scenarios for the origin of the Solar System: (1) haphazard accumulation—the chance hypothesis, (2) catastrophic origin—the sudden hypothesis, and (3) condensation out of a gas cloud—the nebular hypothesis. Each of these has been the favored explanation at one time or another; however most astronomers find the last of these, the nebular hypothesis, the most appealing and internally consistent with the observations listed in the previous section.

The nebular hypothesis advances that the Solar System formed over a relatively brief time as a direct consequence of the Sun's formation from the collapsing interstellar medium—or nebula. In the **solar nebula hypothesis**, our Sun and Solar System originated in a large interstellar cloud consisting primarily of hydrogen and helium with only tiny amounts of other elements in the form of small dust grains. Astronomers routinely observe examples of such clouds around the galaxy and suspect that they are very similar to what ours would have looked like about 4.5 billion years ago. Either because it radiated away enough heat or possibly because of some external trigger like the shock wave from a nearby supernova, the cloud began to collapse onto itself. It is unlikely that the initial cloud could have had exactly zero spin and so, even with initially only a very small tendency to spin in a particular direction, its rate of rotation increased dramatically as the cloud collapsed (the reason is exactly the same as why figure skaters pull in their arms to speed up their spin). The result was a rapidly rotating disk of material with a bulge at the center.

Such a formation process leads directly toward an account of the regularity in the dynamical properties of the Solar System. The current constituents of our Solar System were formed out of a system that, because of the basic physics of rotational motion, was like a thin pancake spinning about its central axis. It was then inevitable that all of the planets would orbit in almost the same plane and rotate in the same direction.

Condensation: Forming the Raw Materials

Directly after the collapse of the solar nebula, the cloud would have still been a fairly homogenous (everywhere the same) collection of gas molecules and small grains of heavier elements. At the core of the solar nebula, a large bulge formed very early and grew under the force of its own self-gravity. Just as a falling hammer picks up speed and energy as it falls to the floor under the influence of the Earth's gravity, the molecules drawn towards the Sun by its enormous gravity picked up energy and were compressed. This process caused the Sun to be hot enough to sustain its nuclear reactions in the same way that all stars form. Further, because of its enormous mass, even the lightest of gas molecules were unable to escape. This left the Sun with much the same chemical composition as its parent nebula.

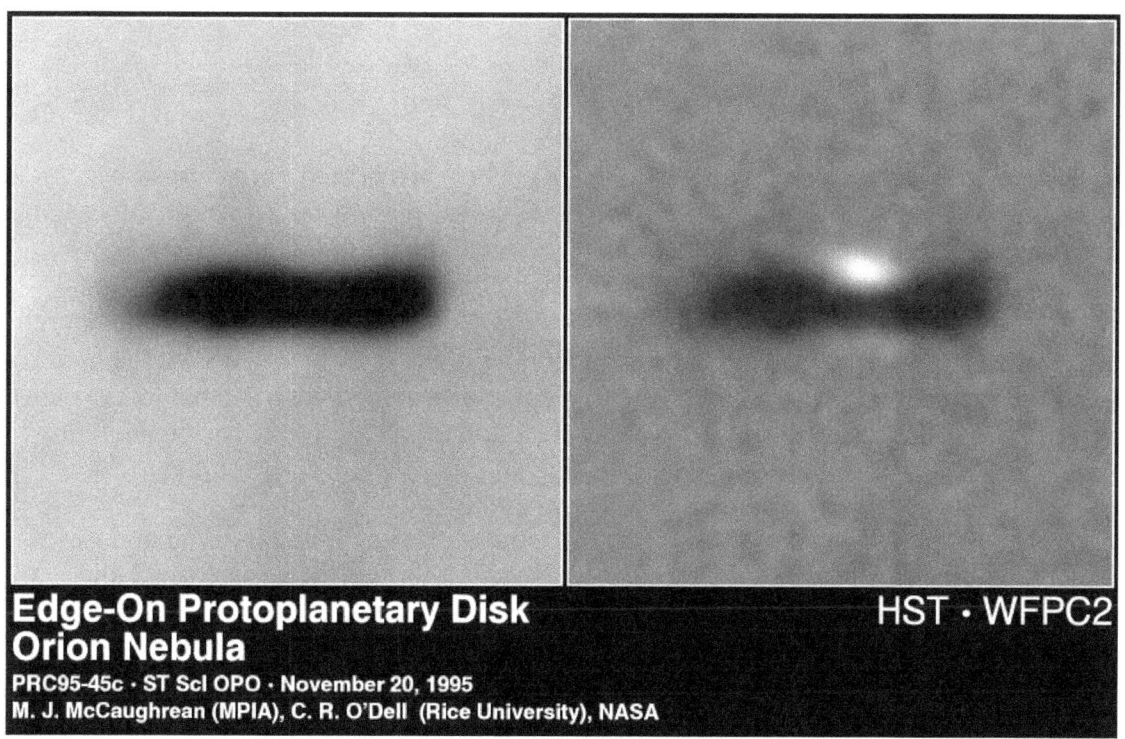

**Edge-On Protoplanetary Disk
Orion Nebula**
PRC95-45c · ST ScI OPO · November 20, 1995
M. J. McCaughrean (MPIA), C. R. O'Dell (Rice University), NASA

HST · WFPC2

The important point in understanding the evolution of the planets is acknowledging the crucial role of temperature. Moving away from the Sun, the temperature decreases with increasing distance. This means that each of the planets evolved in a unique temperature environment. We see an analogous phenomenon everyday when we look at the clouds. As you move away from the surface of the Earth the temperature decreases (usually) and at a certain level it can be cool enough for water vapor to condense into clouds. Thus, the exact nature of the atmosphere depends critically on the temperature, which in turn depends on the distance from the surface of the Earth. This same process of condensation took place in the early history of our Solar System and was similarly regulated by the distance from the heat source. This is the critical idea required to explain why the Sun retained the approximate chemical composition of the nebula while, nearby, the terrestrial planets formed with almost none of the gasses (hydrogen and helium) that characterize the Sun. But, far beyond the terrestrial planets, the gas giants retained the same approximate chemical composition as the original nebula.

In the cooler outer parts of the Solar System, the temperature was low enough that everything condensed—even light elements like hydrogen condensed into small particles and droplets. Since the original material was about 2/3 hydrogen and 1/3 helium, with only bits of everything else, any planet that eventually formed far out in the Solar System would have had the same specific chemical composition.

Closer to the Sun, it would have been much too hot for much of anything in the swirling nebula to condense. Consider first **volatile** elements and compounds—the substances like hydrogen, water vapor, and so on—which boil away quickly. In the inner Solar System, such volatile substances stayed completely gaseous. By contrast, **refractory** elements—heavy elements like nickel and iron—condensed and became solid. Near the hot Sun, therefore, small grains of some materials condensed, consisting of things like metal oxides, alloys of nickel and iron, and so forth. The refractory elements quickly grew to the size of pebbles while the volatile elements remained as gas near the Sun.

Accretion: Planet Construction Begins

Accretion is the name given to the process in which small pebbles, which had been formed by condensation, began to clump together to form larger bodies called **planetesimals**. Since all of the raw materials were orbiting in the same direction, many of the collisions would have been slow enough for the chunks to stick together rather than annihilating one another. So, out of the vast cloud of pebbles grew a collection of planetesimals—most were only 10-100 km across—numbering in the trillions. Not all of these accreted objects would have been the same size and soon the larger ones (called **protoplanets**) attracted the smaller ones by virtue of the large gravitational forces they exerted. The process continued as a few larger objects grew at the expense of the surrounding tiny planetesimals.

Since the amount of raw material near the Sun was small, consisting only of the trace amounts of heavier elements found in the solar nebula, the central planets that formed there were small and rocky. By contrast, those planets evolving in the cooler outer reaches of the Solar System had a great deal more material to work with and thus grew very large. This entire process is thought to have taken only about 100 million years. This might sound like a long time but remember that the Solar System is about 4.5 billions years old. This 100 million-year formation time only represents about 1/45 of its current age. (About 5 months of your life if you are 20 years old.)

It is worth emphasizing that, in the solar nebular hypothesis, the formation of a planetary system is no longer a particularly unlikely event. Therefore, if this hypothesis is correct, it means that we *should expect* to find planetary systems around many neighboring stars—an enticing prospect to say the least.

The preceding description of the evolutionary history of our Solar System does present a coherent conceptual model that explains many of the properties we now observe. However, it would be misleading to suggest that the case is completely solved. There remain a number of details—including the formation of our own Moon—that do not fit neatly within this scheme. Although most astronomers agree on the general formation process described above, the field of Solar System formation and evolution remains an active area of inquiry.

> **Box 13-1: Solar System FAQ**
>
> How did the moons form around planets?
> In most cases, the moons were probably captured planetesimals that fell into the gravitational fields of their parent planets. Of course the Jovian planets were much larger than the terrestrial planets and also evolved in a much richer environment so it is easy to see why they have so many more moons.
>
> What about Pluto?
> We have kind of dismissed Pluto because it does not fit nicely into our classification scheme. It has a relatively low density like the Jovian planets and yet it is small like the terrestrials. In many ways Pluto remains a mystery. In 2006 the International Astronomical Union defined a planet as having the following 3 characteristics: the body orbits the Sun, it is large enough to have a round shape (hydrostatic equilibrium) and it has cleared a path in its orbit. Pluto fails on the last characteristic and it has been joined by several other 'dwarf planets' including Eros and Ceres.
>
> Why is there no planet at 2.8 AU?
> There is a large collection of asteroids between Mars and Jupiter but these evidently were unable to coalesce into a single body. The current understanding is that the gravitational forces of Jupiter were primarily responsible for continually "stirring the soup" and preventing planet formation.
>
> Why is Venus spinning backwards?
> It is likely that Venus experienced an event in which it was struck by a large planetesimal late in its formation and that this impact knocked Venus upside down giving it an apparent reversed spin direction. A similar impact was likely responsible for knocking Uranus on its side.

13.4 Other Planetary Systems?

Today, most astronomers agree that all stars form from the collapse of enormous interstellar clouds of gas. The materials that are left over from one of these collapsed clouds can accrete into planets. If this scenario is true, then there should be millions of planetary systems just like our own.

Astronomers have spent lifetimes trying to build telescopes strong enough to see planets orbiting other stars. This is an enormously difficult task and, to date, *very few* extra-solar planets have actually been *seen*. The problem is that stars are many, many times brighter than any planets around them would ever be. Consequently, planets are hidden in the glow of their parent star. An astronomer working at the closest star to our Sun, a star called Proxima Centauri, would not be able to see any of the planets in our Solar System even with a telescope as fine as the Hubble Space Telescope (HST). Astronomers therefore have to look for indirect evidence.

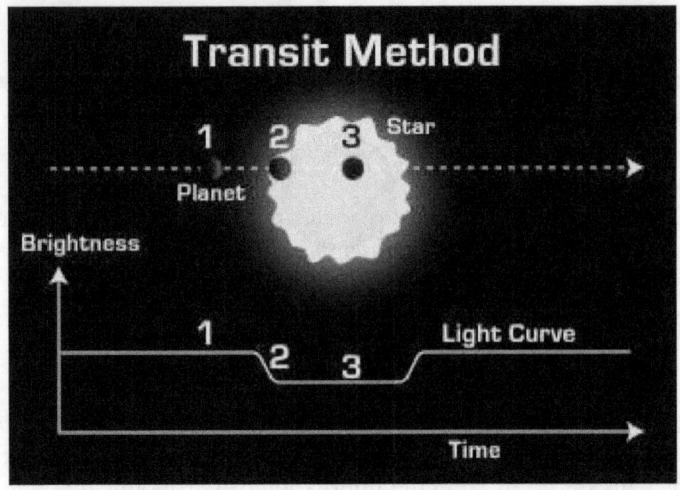

One indirect approach (the **Doppler method**) that astronomers take is to look for "wobbles" in the motion of stars. If a large planet is orbiting a central star, the gravitational interaction between a star and a planet will cause the star to oscillate back and forth a tiny amount. This strategy has resulted in the discovery of dozens of extra-solar planets. There are no pictures of these planets, but we know the planets are there because of how the central star moves.

The other commonly used method to detect extra-solar planets is the **transit method**, in which a telescope watches a planet as it transits or moves in front of its star. As the planet passes in front of its star the total amount of light that the star is giving off is decreased by just a bit, and the telescope can use that to measure how far the planet is from the star and other planetary properties.

Chapter 14
Space Debris

Thus far we have naturally focused our attention on the largest members of the Solar System—those objects that, for the most part, are visible to the naked eye and whose motions have preoccupied astronomers for over two millennia. However, the Solar System is composed of hundreds of thousands, even millions, of smaller bodies in addition to planets and moons. These meteors, asteroids, and comets have mystified sky watchers and dazzled astronomers throughout history. Now we invite you to consider some of these other objects that populate our Solar System and occasionally treat us to unusual and sometimes spectacular displays in the night sky.

14.1 Meteors

Especially in dark locations like Montana, if you spend any time at all out of doors at night you will doubtless see a "shooting star." This brief but bright streak of light results from a small chunk of matter (most no larger than a pea) striking the Earth's atmosphere. The entry of this material heats the surrounding air to several thousand degrees, giving off light. This pebble of dust, prior to hitting Earth's atmosphere is called a **meteoroid**. However, once it collides with the atmosphere, it is known as a **meteor**. If, by some chance, the debris is lucky enough to survive the trip through Earth's atmosphere, it is called a **meteorite**. Astronomers estimate that a meteoroid must be about 1 meter (a yard) across to have any chance of surviving the brimstone trip through Earth's atmosphere.

Meteorites are classified in several different ways. One three-class system uses composition to classify meteorites into *iron meteorites*, *stony meteorites*, and *stony-iron meteorites*. Another system classifies meteorites by how they are recovered: if someone sees a meteorite fall to the ground and picks it up, it is classified as a *fall*; if someone discovers a meteorite in her yard, it is called a *find*. More interesting than the unimaginative names given to these meteorites is their perplexing distribution. Astronomers do not know whether irons, stonys or stony-irons are the most abundant in the Solar System. Almost all meteorites astronomers have are *finds*. These finds are almost entirely iron meteorites. On the other hand, almost all meteorite *falls* are stony meteorites. Part of what makes this complicated is that it is easy to find an iron meteorite in the field—irons usually look very different than any surrounding rocks. However, stony meteorites look like regular rocks and are very difficult to distinguish from Earth rocks. Stony meteorites are rarely found except when first seen falling from the heavens. This is an ongoing mystery. To complicate matters, it has been recently confirmed that some meteorites found in Antarctica—an easy place to find meteorites because of the empty snowfields—have originated from the surfaces of the Moon and even Mars.

Meteorites can contain a wealth of scientific information and, being rare, are highly prized. In fact, given the recent publicity surrounding meteorites as possibly providing evidence for life on Mars, private collectors are now placing estimated values ranging into the millions of dollars for small meteorites. This is making it all the more difficult for scientists whose work depends on access to these objects. Given the rarity of meteorites, it is perhaps surprising to learn that astronomers estimate that the daily yield of space matter raining down on the Earth is in the hundreds of tons. But most of this results from small meteoroids that completely vaporize during entry and therefore fall to Earth as dust. Only a very small fraction of meteoroids are large enough that they slow substantially upon entry and (even though they are heavily vaporized) survive as small pieces.

14.2 Meteor Showers

Any night of the year, you can expect to see meteors at a rate of a few every hour due to the random occurrence of debris along the Earth's orbit around the Sun. However, a couple of times each year, this slow rate can jump to more than 60 meteors per hour, causing many amateur astronomers to forfeit a night's sleep to view the spectacle. These events are called meteor showers and occur when the Earth passes through a particularly dense collection of space debris, thought to be left by a comet passing near the Sun and depositing some of its mass. Even though they are small, these bits of material continue to orbit the Sun along the same path as the parent comet although they tend to spread out. Because the Earth follows the same path around the Sun every year, we pass through these debris clouds at the same time every year. One of the most exciting meteor showers—known as the Perseid shower—occurs each August 12 as the Earth once again encounters debris that is believed to have been deposited by the comet Swift-Tuttle. The table on the previous page lists the major meteor showers of the year and the approximate frequency at which meteors are expected. However, it is important to realize that the comet debris is not uniformly distributed, therefore the rates vary substantially from year to year. Experienced meteor watchers know that the best time to look for meteors is between midnight and dawn. This is because after midnight observers are on the Earth's leading side as it travels around the Sun. Because most meteors occur as a result of the Earth plowing into the space debris—as opposed to meteoroids hitting Earth—there will be more collisions to observe after midnight. This is similar to the difference between walking forwards or backwards during a snowstorm.

The Yearly Meteor Showers

Name	Date	Approx. Hourly Rate	Associated Comet
Aquarid	May 4	5	Halley
Perseid	Aug. 12	40	Swift-Tuttle
Orionids	Oct. 22	13	Halley
Taurid	Nov. 1	5	Encke
Leonid	Nov. 17	6	Tempel 1
Geminid	Dec. 14	55	Ikeya

14.3 Asteroids in the Asteroid Belt

The number sequence of Bode's Rule suggested that there was a planet missing between Mars and Jupiter at 2.8 AU. On January 1, 1801, Sicilian Giuseppe Piazzi found the predicted trans-Martian planet. A year later, it was discovered again on January 1, 1802 by mathematician Karl Fredrick Gauss. Now called Ceres, this object was the first of hundreds of planet-like objects discovered in the region. These objects are now called asteroids—a term coined by William Herschel to distinguish them from the major planets. Only months after Ceres was found came the discovery of Pallas, which was then followed by announcements of Juno and Vesta. Most asteroid discoveries have actually been made by amateur astronomers. Unitarian minister Joel Hastings discovered 39 asteroids between 1905 and 1914.

Most of the asteroids are very small indeed. Ceres is the largest at just under 1000 km across—about the size of Montana. Of the thousands known, only 30 are larger than 200 km—the distance between Bozeman and Billings, Montana. As one could imagine, determining small asteroid diameters from nearly 200 million miles away is an arduous task. The most common measurement technique is to wait for the asteroid to pass in front of a distant star. This asteroid-star eclipse is called an **occultation**. By timing how long an asteroid takes to pass in front of a background star and knowing how fast the asteroid moves, a diameter can be easily calculated. The longer the star is obscured the larger the asteroid. More recently, several spacecraft have passed near asteroids and radioed back beautiful images showing jagged edges and cratered terrain. The Hubble Space Telescope has also substantially enhanced the study of the larger asteroids.

14.4 Earth-Crossing Apollo Asteroids

Similar to the planets, most asteroids are in nearly circular orbits. There are, however, a few rogues, known as the **Apollo asteroids**. The Apollo asteroids have orbits that are highly eccentric and therefore dodge inside Earth's orbit during part of their trip around the Sun. For instance, the asteroid Icarus gets within 0.2 AU of the Sun and passed within 0.04 AU of the Earth in 1968. In 1991, an asteroid estimated to be less than 9 meters in diameter passed within 170,000 kilometers of the Earth—that is less than half of the distance to the Moon!

On average, Earth is struck about once every ten thousand years or so by an asteroid large enough to leave an impressive crater. These collisions are still usually benign events as the Apollo asteroids are small—the Apollo asteroids seem to have a maximum diameter of about one kilometer. However, it is possible for the Earth to be struck by larger asteroids. In fact, an asteroid (or comet) impact is the leading theory to explain the mass extinction (which included dinosaurs) that took place about 65 million years ago.

A 10-kilometer (6-mile) wide asteroid striking the Earth would have released an energy equivalent to 10 million or more of the most powerful hydrogen bombs ever created. This would have released enough dust to occlude the Sun for several years, resulting in a tremendous change in the climate and the extinction of the majority of existing species. This would occur even if the asteroid landed in the oceans—a likely landing place as three-quarters of Earth is covered by water.

14.5 Comets

Probably the most disturbing sky object ever seen in ancient times was a comet. Imagine being completely unable to predict the appearance of fuzzy, star-like objects with glowing tails that slowly, over several months, moved across the sky. Universally, comets were viewed as harbingers of doom across generations and cultures. Some comets are bright enough to be seen during the day and have tails that span halfway across the sky. These ominous comets were once thought to bring tidings of drought, famine, disease, and death. Today, in contrast, astronomers find about a dozen new comets every year and delight in their spectacle. Many of these are found by amateur astronomers who dedicate hundreds of nights every year looking for faint, fuzzy objects that might be a comet that will bear their name.

The anatomy of a comet is simultaneously simple and complex. Most astronomers use a conceptual model attributed to the genius of Fred Whipple in the 1950's. The "dirty snowball" model, officially known as the *icy-conglomerate model*, proposes that comets are a collection of ice and dust packed together. When this snowball approaches the Sun, the Sun's heat and solar radiation liberates material from the surface to form two **tails** and a tenuous atmosphere, called the **coma**, surrounding the **nucleus**. It is this thin coma, reflecting the Sun's light, which we see from Earth. With each pass around the Sun, more and more of the comet disintegrates leaving a dust trail that eventually is the source of the earlier described meteor showers. The gas tail of a comet does not trail behind the comet, rather, it always points directly away from the Sun. This is because it is the Sun's radiation (in the form of the solar wind) that moves ionized gas and dust away from the comet's nucleus. The second tail is made of dust and trails the comet, often in a curved shape.

Box 14-1: Comet Collision

In 1993, astronomers discovered that comet Shoemaker-Levy 9 has been pulled apart by Jupiter's gravity. Even more surprising was the realization that it would impact Jupiter. Never before had the collision of a comet with a planet been observed. Fragments of comet P/Shoemaker-Levy 9 collided with Jupiter on July 16-22, 1994 and the results were spectacular. At least 20 large fragments impacted the planet at 60 kilometers (37 miles) per second, causing plumes thousands of kilometers high. This left great dark scars, which lasted for months after the collision.

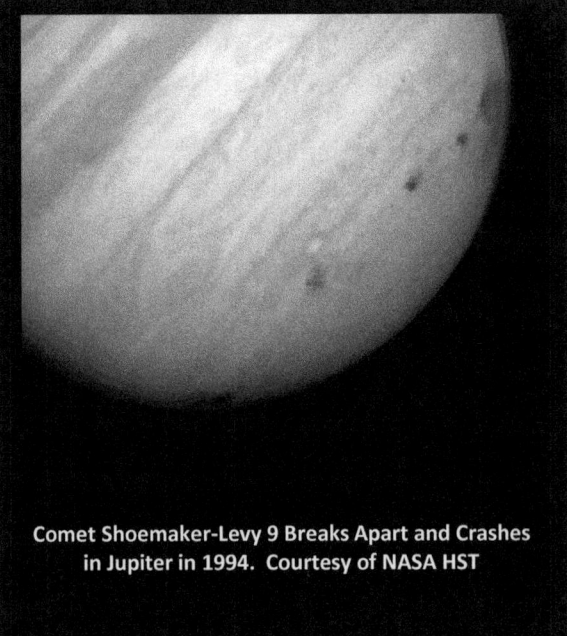

Comet Shoemaker-Levy 9 Breaks Apart and Crashes in Jupiter in 1994. Courtesy of NASA HST

In general, astronomers classify comets into two distinct classes depending upon the time scales over which they reappear back near the Sun. Most comets are **long-period comets**. These comets are in orbits that take them far into the outer Solar System but still within the reach of the Sun's gravitational pull. The other type of comet is called a **short-period comet**. These comets are perhaps much better known because of persistence. They follow Kepler's laws of planetary motion and have highly predictable orbits. For example, Halley's comet returns to the inner Solar System about every 76 years. Clearly, any explanation for the origin of comets must accommodate the existence of both these classes.

In 1949, English astronomer K.E. Edgeworth suggested that short period comets might come from a "comet reservoir" located just beyond Neptune. One year later, Dutch astronomer Jan Oort proposed something that was not dissimilar but was widely accepted. Oort proposed was the existence of an enormous cloud of comets extending out to more than a light year—about 70,000 AU or more than 1500 times the radius of Pluto's orbit. Today this cloud of comets is known as the Oort cloud. He further suggested that the passing of nearby stars causes gravitational disturbances within this cloud that eject comets into a path that carry them near the Sun. Astronomers generally accept that this is indeed the source of long period comets. The existence of such an Oort cloud is also consistent with the nebular hypothesis for the formation of the Solar System. It is now thought that the Oort cloud extends from about 50,000 to 150,000 AU in a giant spherical halo around the Sun.

In 1951, based on the results of theoretical models of Solar System formation, Dutch-American astronomer Gerard Kuiper suggested that the process that led to the formation of Pluto should have extended well beyond Neptune. He further predicted that there should be an entire belt of comet-like objects in the region just beyond the orbit of Pluto which, having formed there, should be confined to a relatively thin disk as with the rest of the planets. Kuiper's work went largely unnoticed for more than 30 years.

Julio Fernandez of Uruguay showed (in 1980) that it was mathematically unlikely that the 150 or so known short-period comets could have originated in the Oort cloud. He convincingly argued that it was much more likely that these objects were originating in orbits just beyond Neptune. Further theoretical work predicted the existence of at least 100 million objects greater than the size of Halley's Comet to reside in what is know known as the Kuiper belt. Thus began the search for "Kuiperoids."

In 1992, the first of these Kuiperoids was found at 41 AU; by 1995 the number had grown to more than 25. In 1995 the Hubble Space Telescope was used to take a series of 34 images of a small piece of sky and, by overlaying the images, the researchers were able to identify 29 new Kuiperoids. The number of objects found was consistent with the number predicted in computer simulations for the very small section of sky observed. This is now considered as definitive evidence for the existence of the Kuiper belt.

Below is a Hubble Space Telescope image that illustrates the difficult detection of objects in the Kuiper Belt. The left image, taken on August 22, 1994, shows the candidate object inside the circle. The right picture, take of the same region one hour forty-five minutes later shows the object has apparently moved in a manner consistent with a Kuiperoid's predicted orbit.

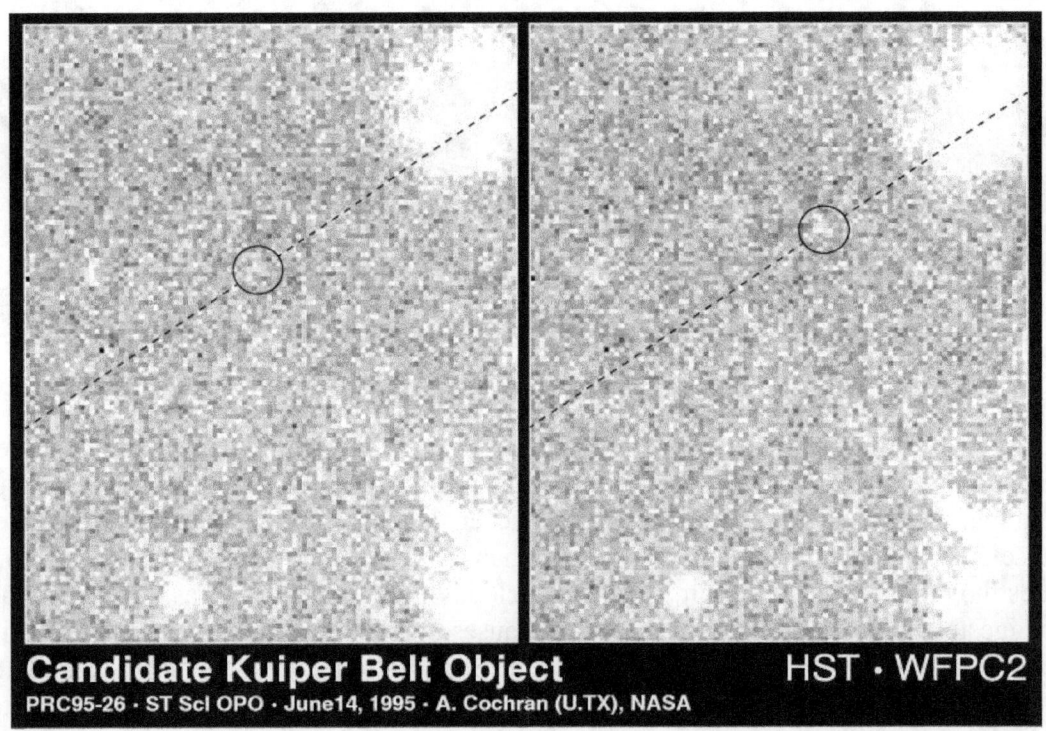

Appendices

In the study of astronomy it is very important to develop a feeling for scales of time and space (distance) that are well outside the domain of everyday experience. The numbers are astronomically large, if you will pardon the pun. Technically, we represent large numbers with the use of powers of ten notation. For instance, we know that the radius of the Earth is about six thousand four hundred kilometers, which we can write as 6.4×10^3 km and read as 6 point 4 times 10 to the power of 3. The simplest way to think about this is that the exponent—the small number written above the 10—indicates how many places to move the decimal point to the left or right. If the exponent is positive the decimal point is moved to the right (i.e., it is a very large number) whereas a negative exponent means that the decimal point is moved to the left (i.e., it is a very small number). With this system we have no problem representing the enormous mass of the Sun as 1.99×10^{30} kg or the tiny mass of a single electron as 9.11×10^{-31} kg. But, being able to represent a quantity is not the same as being able to conceptualize its meaning.

To measure distances astronomers typically use one of three different scales. For measuring distances in the Solar System, the standard ruler is the average Earth-Sun distance, which is about 93 million miles (93×10^6 miles). This distance is called an **astronomical unit (AU)** and is a convenient unit for describing planetary orbits. For instance, the radius of Earth's orbit is 1 AU, the radius of Jupiter's orbit is about 5 AU and the radius of the Pluto's orbit—the outermost planet—is about 40 AU. Even though (in astronomical terms) the AU is a very short distance, it is unimaginably large in human terms. Traveling at a speed of 186,000 miles per second, light takes over 8 minutes to travel 1 AU.

This last example hints at the second ruler that astronomers use for describing large distances. Just like we might state the distance to a nearby town in units of time—"it's about 2 hours to Billings"—astronomers often quote distances in units of time. And, in the same way that we would assume common highway speeds to get to Billings, astronomers also use a common speed: the speed of light. Light travels at a constant rate of 186,000 miles per second through empty space. The distance that light travels in a minute is called a light-minute; the Earth is about 8 light-minutes from the Sun. However, the more common unit is the distance that light travels in a year, which is called the **light-year** (ly). The nearest star to Earth is, of course, our own Sun. The next nearest is a star called Proxima Centauri, which is more than 4 ly from Earth! This means that a radio signal, which travels at the speed of light, takes 4 years to get to Proxima Centauri.

The third common unit that astronomers typically use for describing the distances to stars is called the **parsec** (pc), which is a contraction of parallax-second. A parsec is a little more than three light-years. The details are described in the unit on star magnitudes and stellar parallax.

Although these distances are perhaps unimaginable in human terms, it can be helpful to remember a few relative distances and times to serve as benchmarks. In terms of length scales, the only absolute value that you should remember is that the radius of the Earth is about six thousand four hundred kilometers (i.e., $R_{Earth}=6400$ km). The others we can just remember relative to each other (all are approximate).

Some basic distance scales that you should remember:

- Radius of the Moon: 1/4 R_{Earth}
- Radius of the Moon's orbit: 60 R_{Earth}

- Radius of the Sun: R_{Sun} = 100 R_{Earth}
- Radius of Earth's orbit: 200 R_{Sun} (or 1 AU = 200 R_{Sun})
- Radius of Pluto's orbit: 40 AU
- Distance to nearest star: 250,000 AU

Some basic time scales that you should remember:

- Age of the Universe: about 15 billion years (15 x 10^9 years)
- Age of Earth: about 4.5 billion years
- First life on Earth: about 3.8 billion years ago
- Last great mass extinction: about 65 million years ago
- Emergence of Homo sapiens: about 350 thousand years ago

Probably the most well-known night-sky object is the Moon. Much like the Sun, the Moon rises in the East and sets in the West. However, the phenomenon that has attracted the most attention over history is that the Moon also changes its appearance. Over the course of its monthly cycle, the Moon appears in several shapes, changing from a thin, slender crescent to a round and full Moon and then back to a slender crescent. As viewed from Earth's surface, these varied faces are called **phases**.

Watching the Moon at Sunset

Looking up to the evening sky night after night, an observer facing south would see the Moon rise in the East, move through the sky, and set in the West—but not quite as fast as the Sun moves through the sky. On the first night of the lunar cycle, the Sun and the Moon set at the same time. However, on the next night, the Moon sets about an hour after the Sun does. The Moon appears as a thin, slender **crescent** quite close to the western horizon. On the third night, the Moon, falling further behind the quickly moving Sun, sets about two hours after the Sun. The crescent appears slightly thicker than the night before and is located in a southwest direction. On the fourth night of the lunar cycle, the Moon sets more than 3 hours after the Sun and appears much thicker than the previous night. This change from a thin appearance to a thicker appearance and the accompanying eastward changing position of the Moon relative to the Sun is called **waxing**. It occurs for the first two weeks of the 28-day lunar cycle.

Seven days into the lunar cycle, the Moon appears above the southern horizon at sunset (it is actually quite high in the sky but we refer to this as south). It is called a **first quarter Moon** because it is ¼ into its 28 day lunar cycle. It appears half lit and half dark being round on the right and almost flat on the left. Because it is in the south while the setting Sun is in the west, it will set about 6 hours—about a quarter day—after the Sun sets.

On days 8 through 13 of the lunar cycle, the Moon appears round on the right side and oblong on the left side. This "humpbacked" Moon is called a **waxing gibbous Moon**. It sets between 6 and 12 hours after the Sun does and progressively appears farther and farther toward the eastern horizon each evening at sunset.

> **Box A2.1: Lunar Superstitions**
>
> Superstitions can be defined as a belief, conception, act, or practice resulting from ignorance, unreasoning fear of the unknown or mysterious morbid scrupulosity, trust in magic or chance, or a false conception of causation. Most Moon superstitions relate to the waxing (growing) and waning (shrinking) of the Moon. The Moon is waxing from the time of the new Moon to brightly shining full Moon—about fourteen days. After the full Moon, less and less of the lit side of the Moon is visible until it disappears as a new Moon again. Most of the old superstitions of the changing Moon come from agricultural or human activities. Crops should be planted during the waxing phases whereas crops should be harvested during waning phases. This does not apply to underground crops, such as carrots, which should be harvested during waxing and planted during waning. Also during waning phases, fruit should be picked and firewood should be cut to prevent early rotting. Superstition also dictates that hair should be cut during the waning phase or else it will grow back too fast. And, watch for this one, reading by Moonlight is purported to cause lunacy.

Exactly half way through the lunar cycle—after about 14 days—the Moon rises in the east as the Sun sets in the west. At this point, an observer standing on Earth sees a Moon that is full and round. Of course, this phase is called **full Moon**. The full Moon appears opposite in the sky from the Sun;

the full Moon rises as the Sun sets, it is high overhead at midnight, and it sets in the west as the rising Sun appears in the east.

After the brilliant full Moon, the Moon continues to change; but now it begins to get thinner and thinner with the left side remaining round and the right side withering away night after night. (Left and right designations refer to when the Moon is observed in the south.) This two-week process of thinning is called **waning**. The Moon does not rise until after sunset and rises later and later every night. The phase viewed during days 15-20 is called **waning gibbous**. The waning gibbous Moon looks much like a waxing gibbous Moon except it rises late in the evening and it is the left side that appears round.

Twenty-one days—or ¾ of 28—into the cycle, the Moon rises in the east at midnight. It appears half lit being round on the left and flat on the right. This is called a **third quarter Moon**. Because the Moon takes about 12 hours to rise and set, the Moon does not set until noon the next day. Finally, rising later and later, the Moon is rising just before the Sun rises. This last phase of the Moon is called a **waning crescent** and is most easily seen during the early morning. At the conclusion of the cycle, the Moon once again rises and sets with the Sun—this aligned phase, which is invisible to observers on Earth, is called the **new Moon**. It is invisible because only the opposite side of the Moon is lit by the Sun.

Box A2-2: Some Full Moon Names

Month	Name
January	Old Moon
February	Snow Moon
March	Sap Moon
April	Egg Moon
May	Milk Moon
June	Honey Moon
July	Hay Moon
August	Grain Moon
September	Fruit Moon
October	Hunter's Moon
November	Beaver Moon
December	Yule Moon
2nd Full Moon of a Season	Blue Moon
Full Moon Closest To 1st Day of Spring	Grass Moon
Full Moon Closest To 1st Day of Autumn	Harvest Moon

Watching the Moon from Above the Solar System

From the outside perspective of looking down on the Earth and Moon from above the Solar System, we observe two things. One is that the Moon orbits the Earth once in a little more than 28 days while the Earth follows its annual path around the Sun. The other observation we make from above is that one-half of the Moon is always lit by the sun and one-half is always dark—just like the Earth. As the Earth rapidly spins counter-clockwise, an observer on Earth sees the Sun and Moon rise in the east and set in the west.

As the half-lit, half-dark Moon makes its slow path around the more rapidly spinning Earth, an observer on Earth sees more and more of the lit half of the Moon. When the Earth is situated between the Moon and the Sun, the Moon appears full and bright to the Earth-bound observer. As the Moon comes around to the other side of the Earth, an observer sees less and less on the lit half on the Moon. Finally, when the Moon moves between the Earth and Sun, the lit half of the Moon is facing away from observers on Earth and no Moon is visible from Earth—this is new Moon. It is the spinning Earth that makes the Sun and Moon rise and set whereas it is the orbital path of the Moon that causes the slowly changing Moon phases.

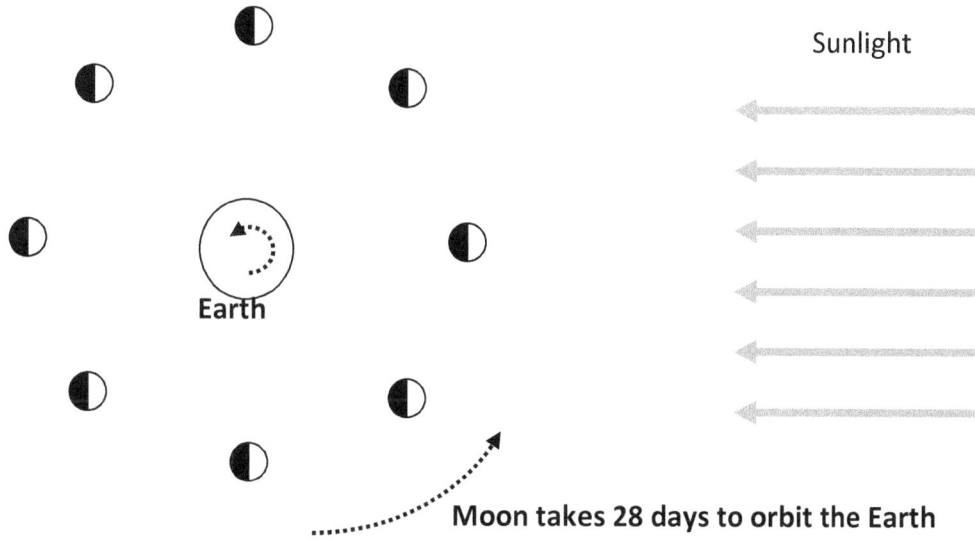

As the principal timekeeper for generations, the Sun rises dependably every morning, journeys across the sky, and sets toward the west. Except for the occasion of a rare solar eclipse, most people pay little attention to this bright sphere in the sky, wholly taking for granted its constancy of brilliance as well as its dependable daily movement.

"Give me the splendid silent Sun, with all his beams full dazzling."

-- Walt Whitman

Over the millennia, the beauty and seeming constancy of the Sun, an object more than 100 times the size of our Earth, have captured the imagination of artists and poets. Today we know that our Sun is but one of 400 billion stars that make up our Milky Way Galaxy. More careful and systematic observations, even before Galileo, reveal that the Sun is not constant; the granulated visible surface is blemished with spots that march across the disk, their numbers waxing and waning with the years.

The Interior of the Sun

Theorists have developed a generally accepted conceptual picture of the Sun's interior. The first clue to the Sun's interior was that the only energy source strong enough to make the Sun shine as bright as it does, (and for as long as it has) is the nuclear fusion of hydrogen into helium. This process exists only at very high temperatures and pressures. This gives us some insight into the conditions at the solar core. The other information we have about the interior of the Sun is gathered in the same way that scientists study the interior of the Earth. The Sun experiences sun-quakes much like Earth experiences earthquakes. By carefully monitoring the way sun-quake waves move through the Sun, helio-seismologists have developed a conceptual model of the Sun interior. This model includes a 16 million Kelvin core of nuclear fusion, a large conduction region where heat moves upwards, and finally a thin convection shell where hot material bubbles to the top, forms granules on the visible surface, cools, and sinks back into the Sun.

Image Courtesy of NASA and Yohkoh

The Photosphere

The part of the Sun that is visible to our eyes is called the photosphere. As shown in the Yohkoh satellite image of the Sun's photosphere on the previous page, daily observations reveal the Sun to be dotted with sunspots. The observed motion of the sunspots across the disk of the Sun shows the Sun to be spinning about its axis about once every 26 days. Spectroscopic analysis of these spots suggests two surprising things. First, sunspots appear dark because they are only 3500 Kelvins—much cooler than the surrounding 5700 Kelvin photosphere. Second, they are regions of intense magnetic fields, which can be thousands of times stronger than the Earth's magnetic field. These magnetic fields push hot gas around on the Sun's surface causing temperature variations. This is a critical aspect of the solar cycles described below.

Eclipse Image of Corona Courtesy of NASA/HAO/NCAR

The Chromosphere and Corona

Above the photosphere, the Sun has a diffuse atmosphere of hot, ionized gas. During a solar eclipse, the bright photosphere is blocked out by the Moon (see figure at top right) allowing the much fainter upper parts of the Sun to be seen. The chromosphere is a thin red layer directly above the photosphere. Directly above the chromosphere and reaching out more than one solar radius is the mysterious corona. With the exception of the glimpses gained during infrequent solar eclipses, early astronomers knew very little about the chromosphere and corona. Today, astronomers are able to create false eclipses by inserting a small metal disk into their telescopes allowing the corona to be studied without waiting for an infrequent solar eclipse. This device is called a **coronograph**. Studies with the coronograph reveal a startling fact. Instead of cooling as gas gets farther from the 5700 K photosphere, gas in the chromosphere reaches 20,000 K and gas in the corona reaches millions of Kelvins! Although there are some tentative theories, this increase of temperature with distance is a haunting mystery to solar scientists.

It is in the observation of the corona that we see some of the most violent eruptions in our Solar System. Occasionally, dynamic twists in the Sun's magnetic field capture high temperature gas and eject it into outer space. These ejections from the corona, called coronal mass ejections or CME's, can contain as much mass as Mt. Everest and move at speeds upwards of a million miles per hour. When directed away from the Earth, they make beautiful displays as shown in the Yohkoh satellite images on the bottom of the

Image Courtesy of NASA and Yohkoh

opposite page. On the other hand, when these CME's are directed at the Earth the results can be devastating. Satellite electronics can be disabled, power companies can lose electricity generation, damaging electrical pulses can be created in long oil pipelines and communication cables, ham radio communications can be disrupted, and astronauts working outside the Space Shuttle can be exposed to dangerous radiation. These CMEs can also result in the glorious light displays known as the aurora borealis—the northern lights and the australis borealis or southern lights.

Solar Cycles

In addition to daily changes, the Sun also undergoes a long-term cycle characterized by sunspots. The average number of sunspots increases, decreases, and then increases again every eleven years.

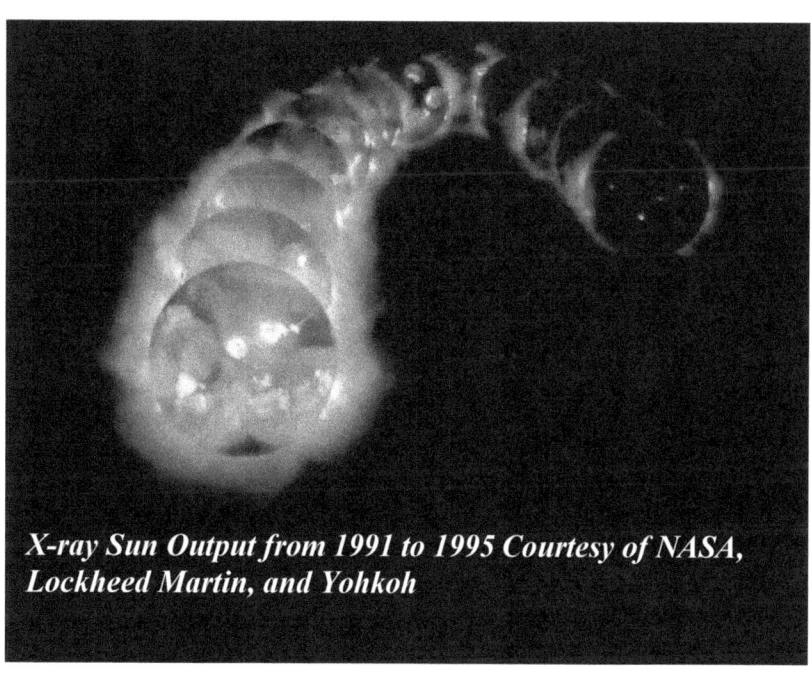

X-ray Sun Output from 1991 to 1995 Courtesy of NASA, Lockheed Martin, and Yohkoh

In 1991, there were hundreds of sunspots visible. By 1996, there were almost no sunspots to be found. By monitoring the changing Sun over hundreds of years, astronomers can now reliably anticipate that the next solar maximum of sunspots will occur in about 2012. Remembering that sunspots are also regions of intense magnetic fields that move hot gas from one place to another, it should come as no surprise that the years leading to 2012 will result in more sunspots, more active regions on the Sun, more CME's, and more solar activity in general. Scientists are aggressively trying to learn how to better predict CME's so that satellite operators, power companies, and astronauts can better prepare for solar effects on Earth.

Dynamical Properties of the Planets

Planet	Average Orbital Radius (AU)	Orbital Period (Earth years)	Orbital Eccentricity	Orbital Tilt (degrees)*	Rotation Time (days)**	Tilt of Spin Axis (degrees)
Mercury	0.39	0.24	0.206	7	58.6	7
Venus	0.72	0.62	0.007	3.39	-243	177.4
Earth	1	1	0.017	0	0.9973	23.45
Mars	1.52	1.88	0.093	1.85	1.026	23.98
Jupiter	5.2	11.86	0.048	1.31	0.41	3.08
Saturn	9.54	29.46	0.056	2.49	0.43	26.73
Uranus	19.19	84.01	0.046	0.77	-0.69	97.92
Neptune	30.06	164.8	0.01	1.77	0.72	29.6
Pluto	39.53	248.6	0.248	17.15	-6.387	118

*This is the tilt of the plane of the planet's orbit compared to the plane of Earth's orbit.

**A negative value is used to indicate a planet that spins in the opposite direction to the Earth.

Physical Properties of the Planets

Planet	Mass (Earth=1)	Radius (Earth=1)	Surface Gravity (Earth=1)	Surface Temperature (degrees Kelvin)	Number of Known Moons
Mercury	0.055	0.38	0.38	100 to 700	0
Venus	0.82	0.95	0.9	730	0
Earth	1	1	1	290	1
Mars	0.11	0.53	0.38	180 to 270	2
Jupiter	317.9	11.21	2.53	124	18
Saturn	95.18	9.45	1.07	97	22
Uranus	14.54	4.01	0.9	58	18
Neptune	17.13	3.88	1.14	59	8
Pluto	0.0025	0.18	0.07	40 to 60	1

Appendix E: Properties of the Thirty Nearest Stars

Star	Spectral Type A*	Spectral Type B*	Distance (pc)	Apparent Magnitude A	Apparent Magnitude B	Absolute Magnitude A	Absolute Magnitude B	Luminosity (Sun=1) A	Luminosity (Sun=1) B
Sun	G2V			-26.72		4.85		1	
Proxima Centauri	M5		1.3	11.05		15.5		0.00006	
Alpha Centauri	G2V	K0V	1.33	-0.01	1.33	4.4	5.7	1.6	0.45
Barnard's Star	M5V		1.83	9.54		13.2		0.00045	
Wolf 359	M8V		2.38	13.53		16.7		0.00002	
Lalande 21185	M2V		2.52	7.5		10.5		0.0055	
UV Ceti	M6V	M6V	2.58	12.52	13.02	15.5	16	0.00006	0.00004
Sirius	A1V	wd**	2.65	-1.46	8.3	1.4	11.2	23.5	0.003
Ross 154	M5V		2.9	10.45		13.3		0.00048	
Ross 248	M6V		3.18	12.29		14.8		0.00011	
Erodani	K2V		3.3	3.73		6.1		0.3	
Ross 128	M5V		3.36	11.1		13.5		0.00036	
61 Cygni	K5V	K7V	3.4	5.22	6.03	7.6	8.4	0.082	0.039
Indi	K5V		3.44	4.68		7		0.14	
Grm 34	M1V	M6V	3.45	8.08	11.06	10.4	13.4	0.0061	0.00039
Luyten 789-6	M6V		3.45	12.18		14.6		0.00014	
Procyon	F51V	wd**	3.51	0.37	10.7	2.6	13	7.65	0.00055
2398	M4V	M5V	3.55	8.9	9.69	11.2	11.9	0.003	0.0015
Lacaille 9352	M2V		3.58	7.35		9.6		0.13	
G51-15	MV		3.6	14.81		17		0.00001	
Tau Ceti	G8V		3.62	3.5		5.7		0.45	
BD+5 1668	M5		3.77	9.82		11.9		0.0015	
Lacaille 8760	M0V		3.86	6.66		8.7		0.028	

* A and B indicate the two stars in a binary system.
**wd stands for white dwarf.

Appendix F: Properties of the Twenty Brightest Stars

Star	Spectral Type		Distance (pc)	Apparent Magnitude		Absolute Magnitude		Luminosity (Sun=1)	
	A*	B*		A	B	A	B	A	B
Sirius	A1V	wd**	2.7	-1.46	8.7	1.4	11.6	23.5	0.003
Canopus	F0Ib-II		30	-0.72		-3.1		1510	
Rigel Kentaurus	G2V	K0V	1.3	-0.01	1.3	4.4	5.7	1.56	0.046
Arcturus	K2III		11	-0.06		-0.3		115	
Vega	A0V		8	0.04		0.5		55	
Capella	GIII	MIV	14	0.05	10.2	-0.7	9.5	166	0.01
Rigel	B8Ia	B9	250	0.14	6.6	-6.8	-0.4	4.6×10^4	126
Procyon	F5IV-V	wd**	3.5	0.37	10.7	2.6	13	7.7	0.0006
Betelgeuse	M2Iab		150	0.41		-5.5		1.4×10^4	
Achernar	B5V		20	0.51		-1		219	
Hadar	B1III	?	90	0.63	4	-4.1	-0.8	3800	182
Altair	A7IV-V		5.1	0.77		2.2		11.5	
Acrux	B1IV	B3	120	1.39	1.9	-4	-3.5	3470	2190
Aldebaran	K5III	M2V	16	0.86	13	-0.2	12	105	0.0014
Spica	B1V		80	0.91		-3.6		2400	
Antares	MIIb	B4V	120	0.92	5.1	-4.5	-0.3	5500	115
Pollux	K0III		12	1.16		0.8		41.7	
Formalhaut	A3V		7	1.19	6.5	2	7.3	13.8	0.1
Deneb	A2Ia		430	1.26		-6.9		5.0×10^4	
Mimosa	B1IV		150	1.28		-4.6		6030	

* A and B indicate the two stars in a binary system.
**wd stands for white dwarf.

120

Appendix G: Monthly Sky Maps

January 1, 10 pm at 45°N

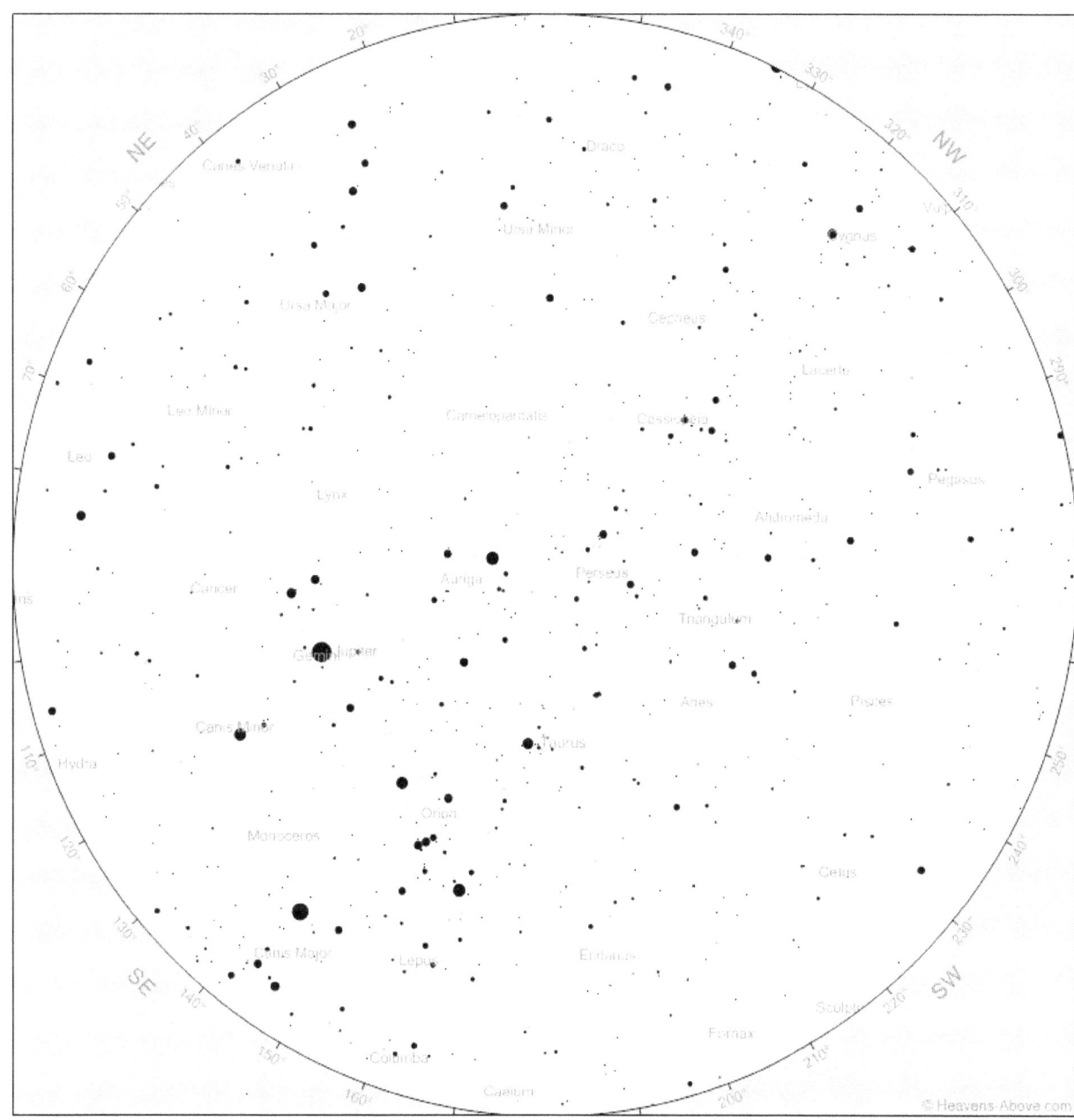

February 1, 10 pm at 45°N

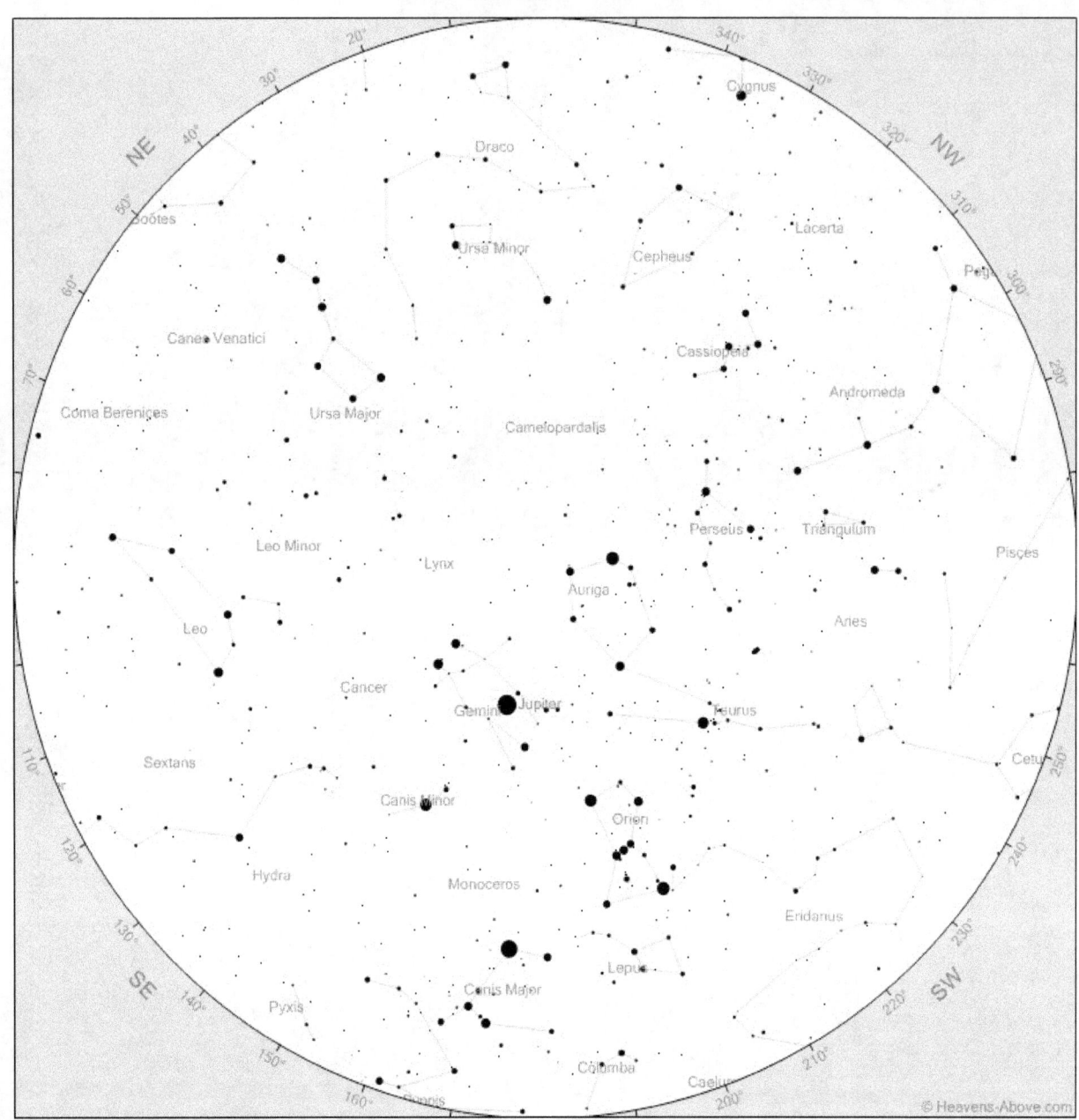

March 1, 10 pm at 45°N

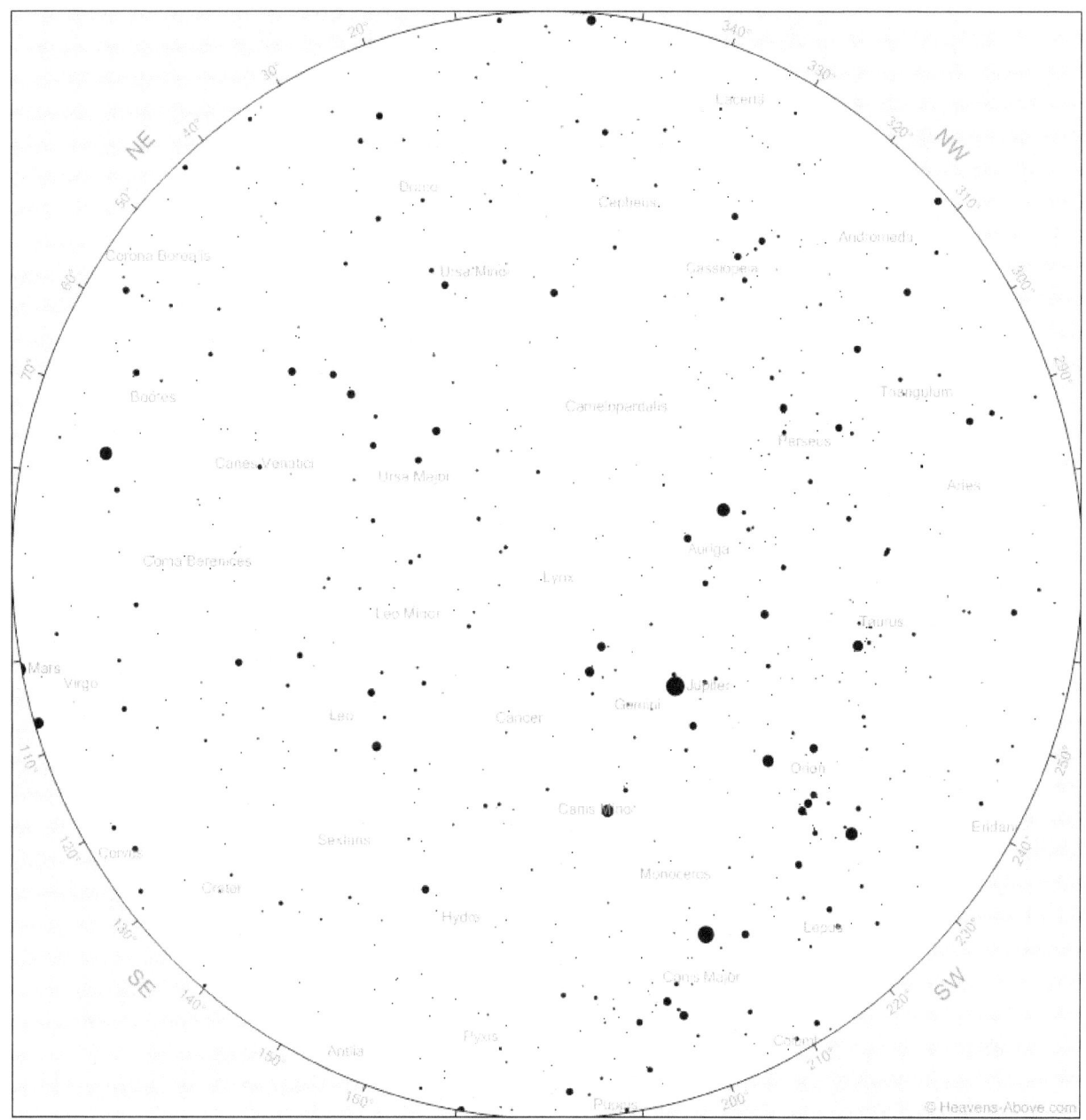

April 1st, 10 pm at 45°N

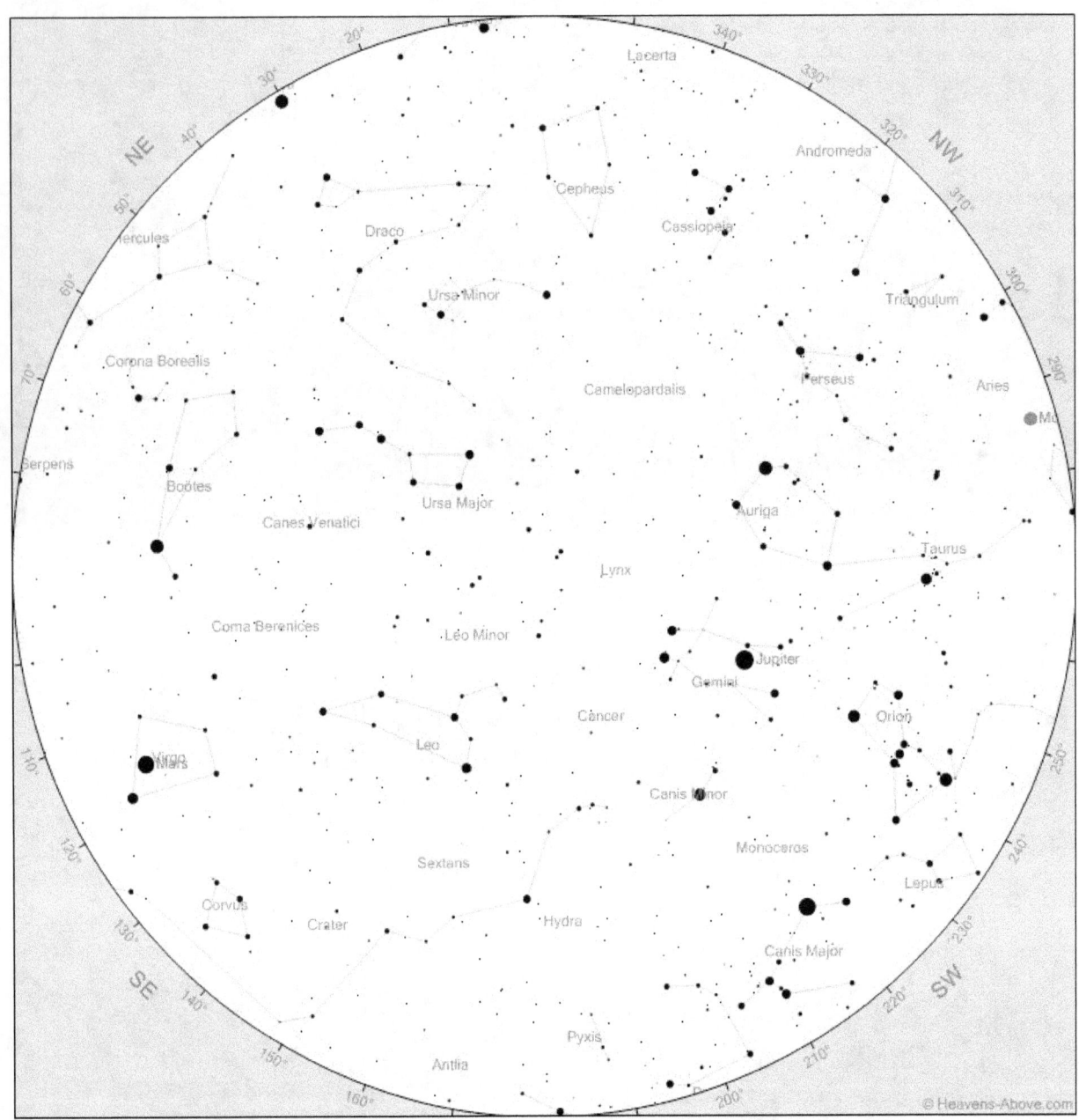

May 1st, 10 pm at 45°N

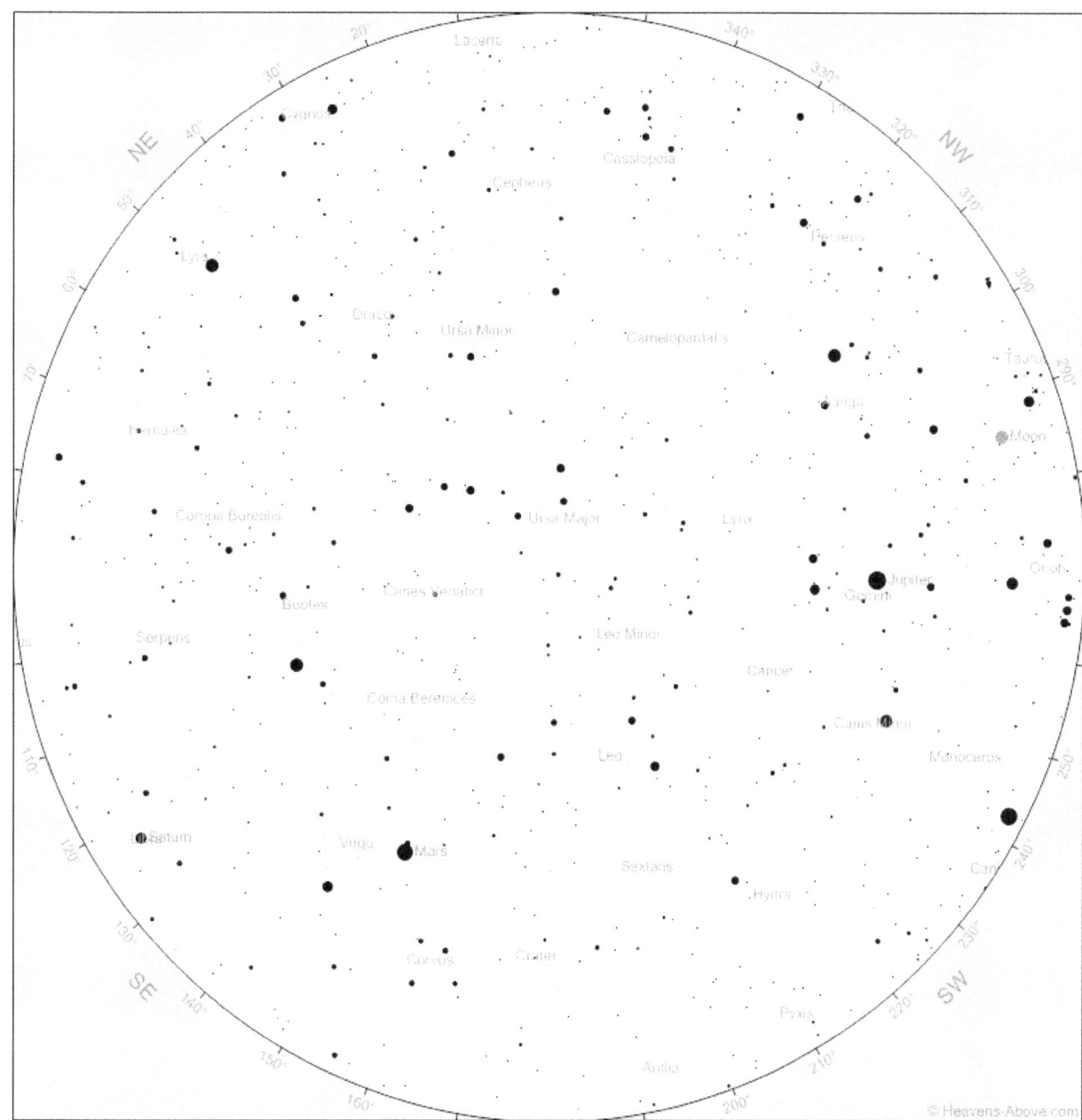

June 1st, 10 pm at 45°N

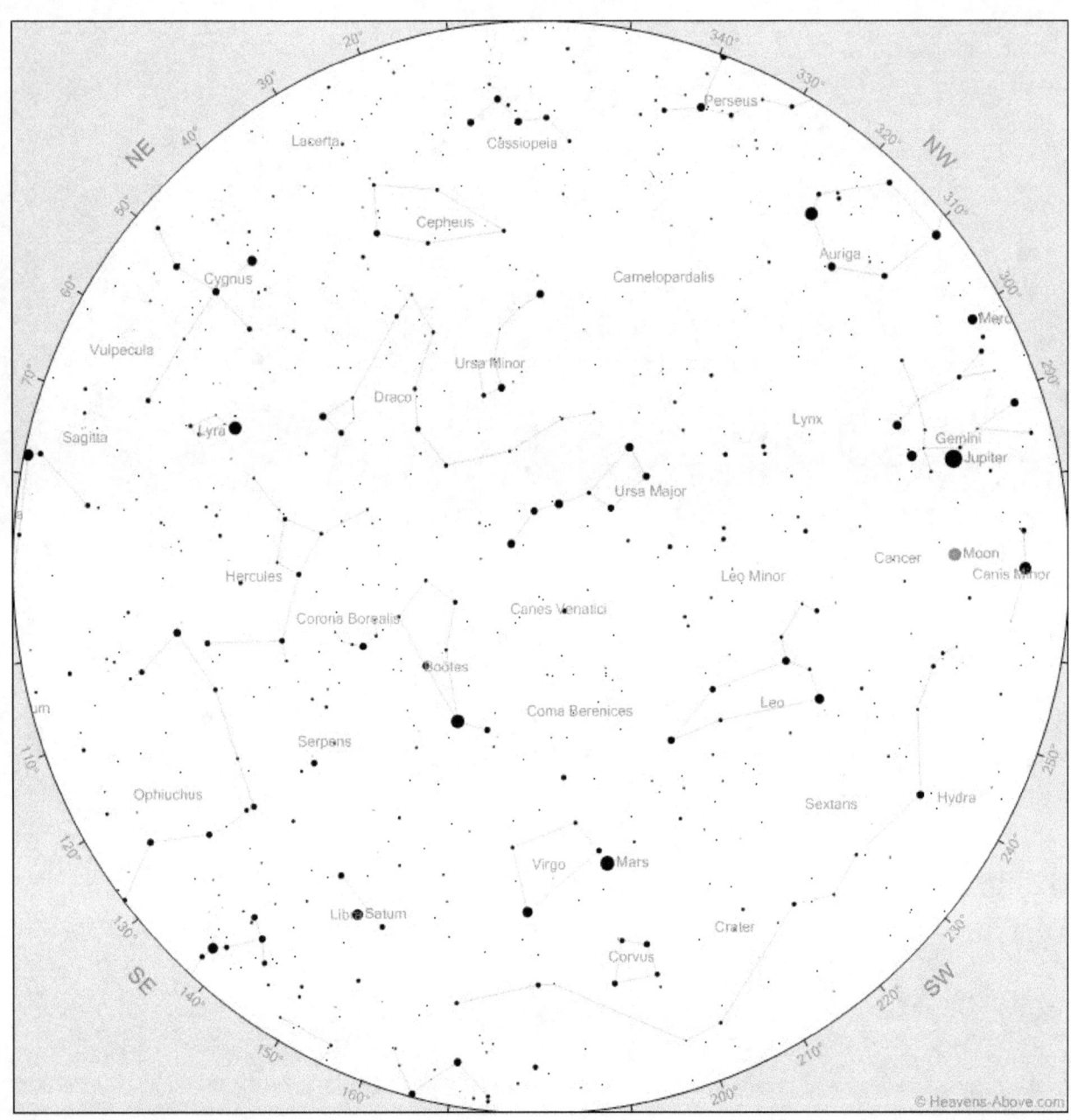

July 1st, 10 pm at 45°N

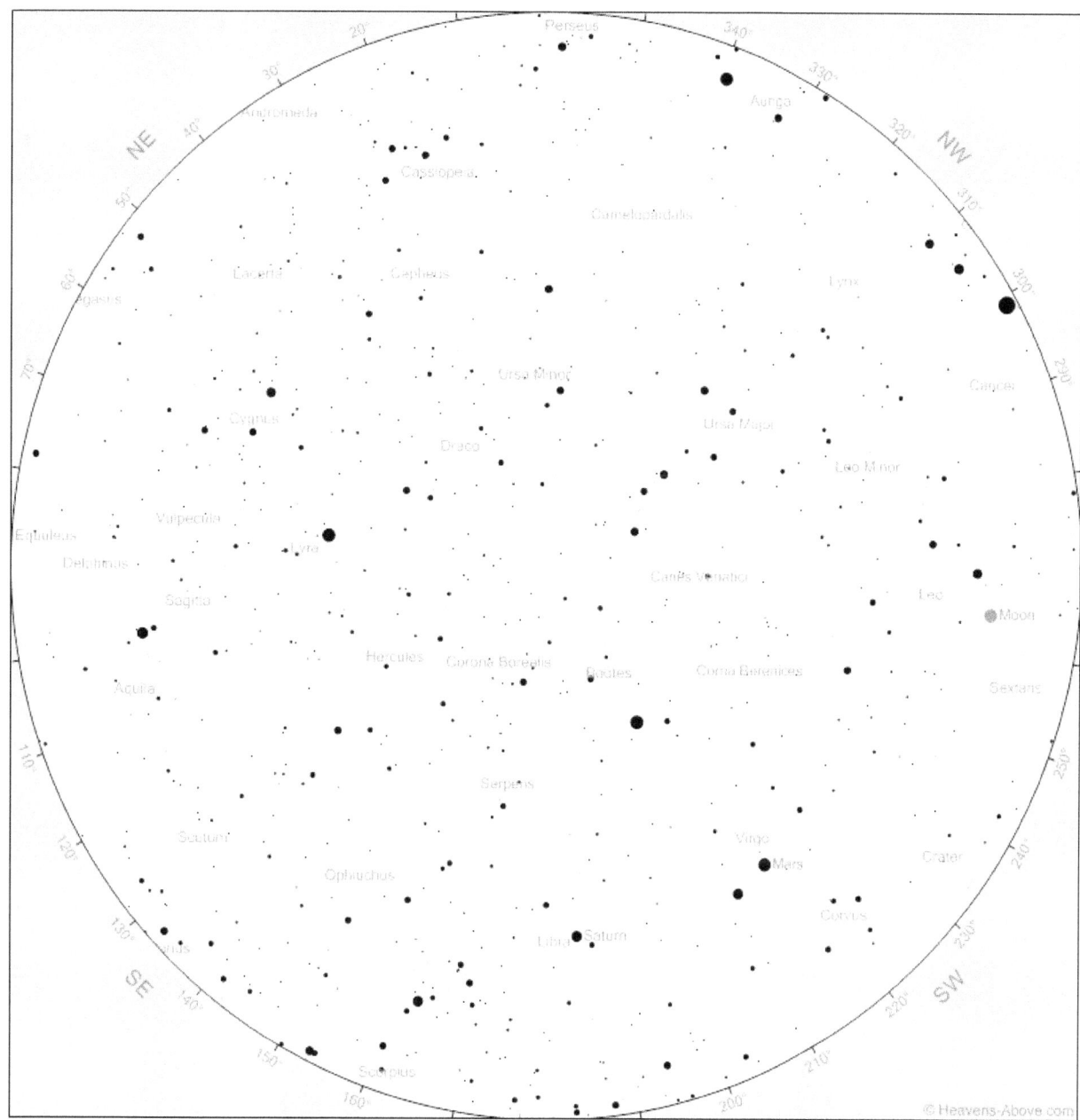

August 1st, 10 pm at 45°N

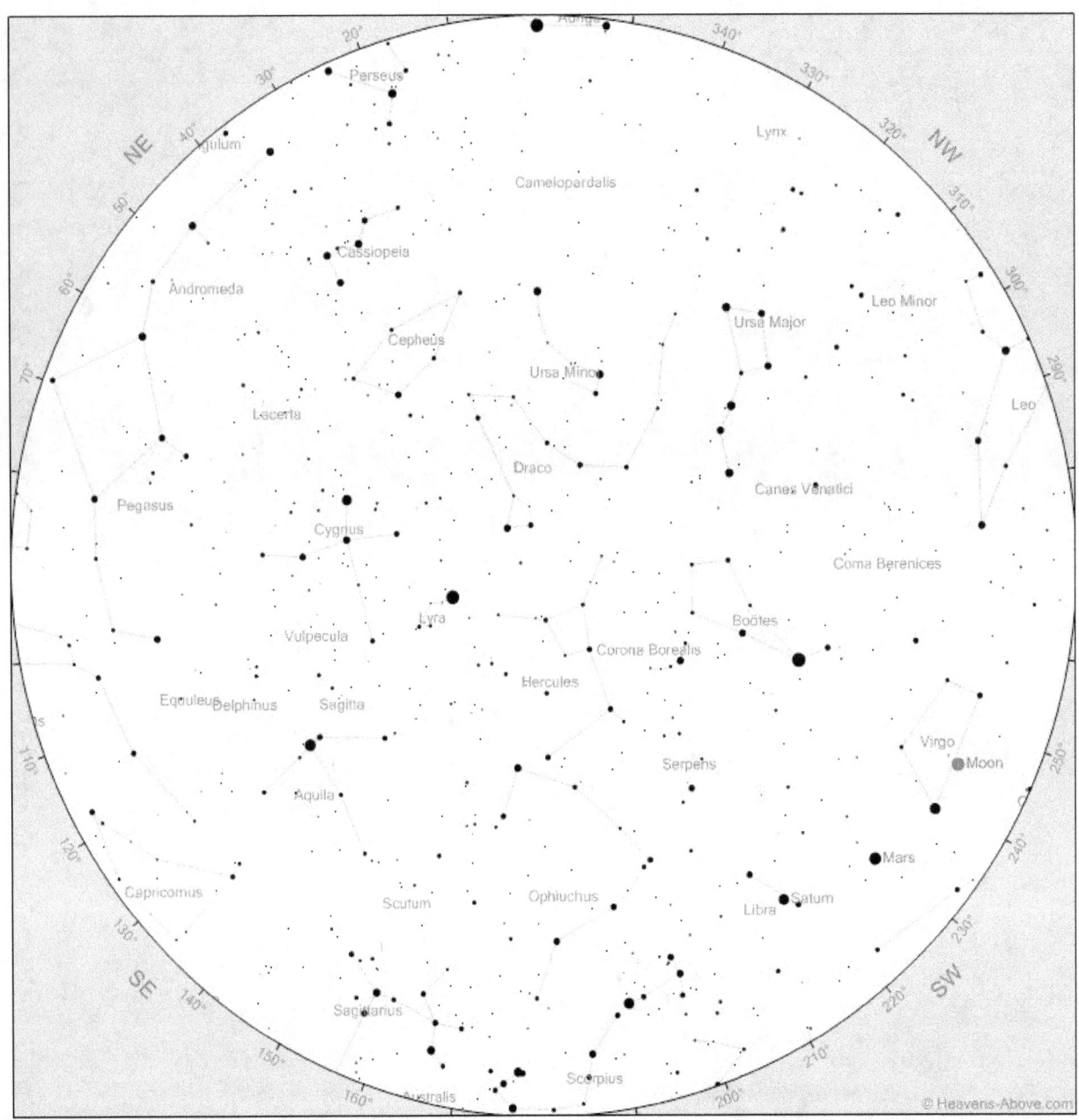

September 1st, 10 pm at 45°N

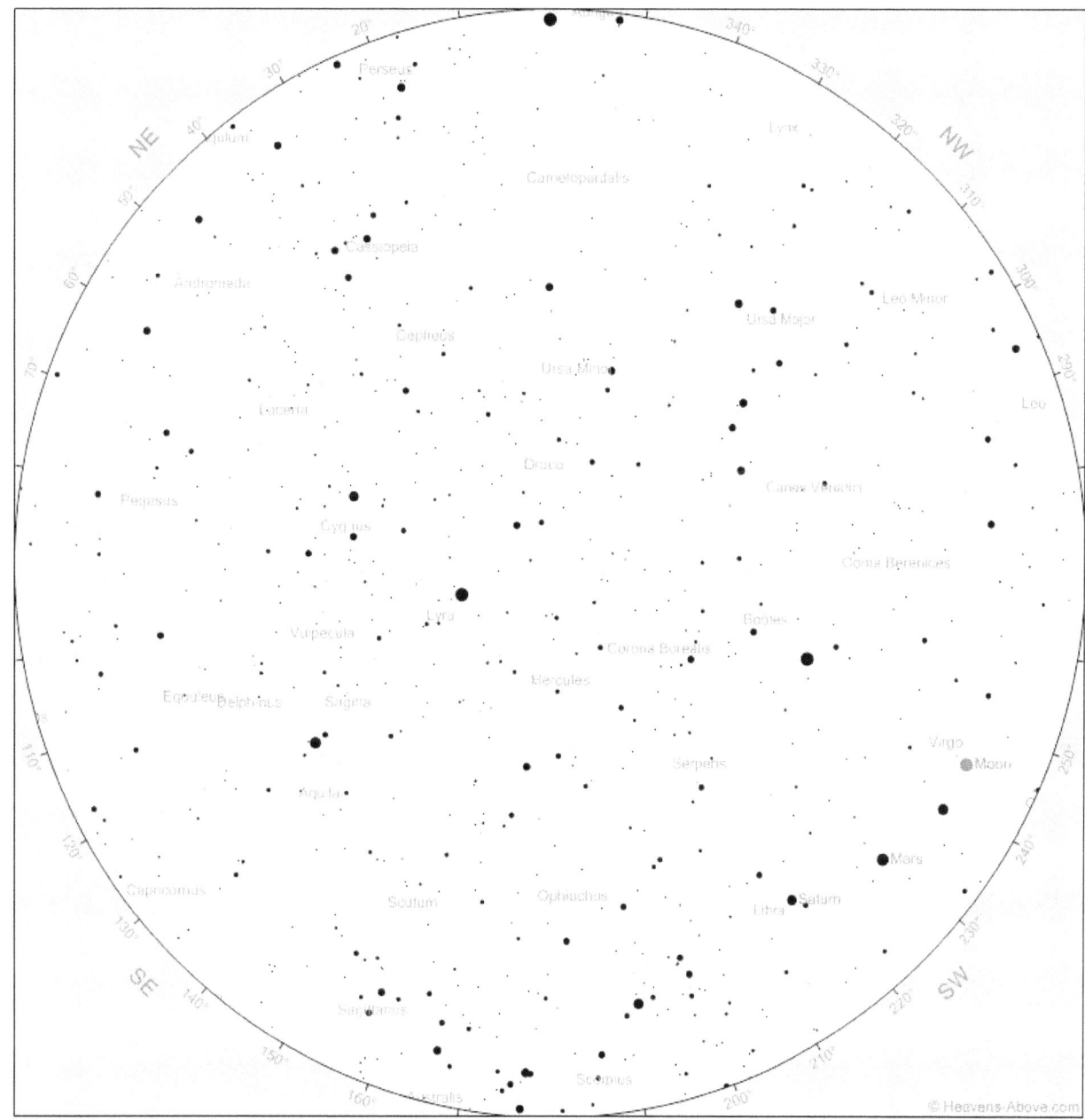

September 1st, 10 pm at 45°N

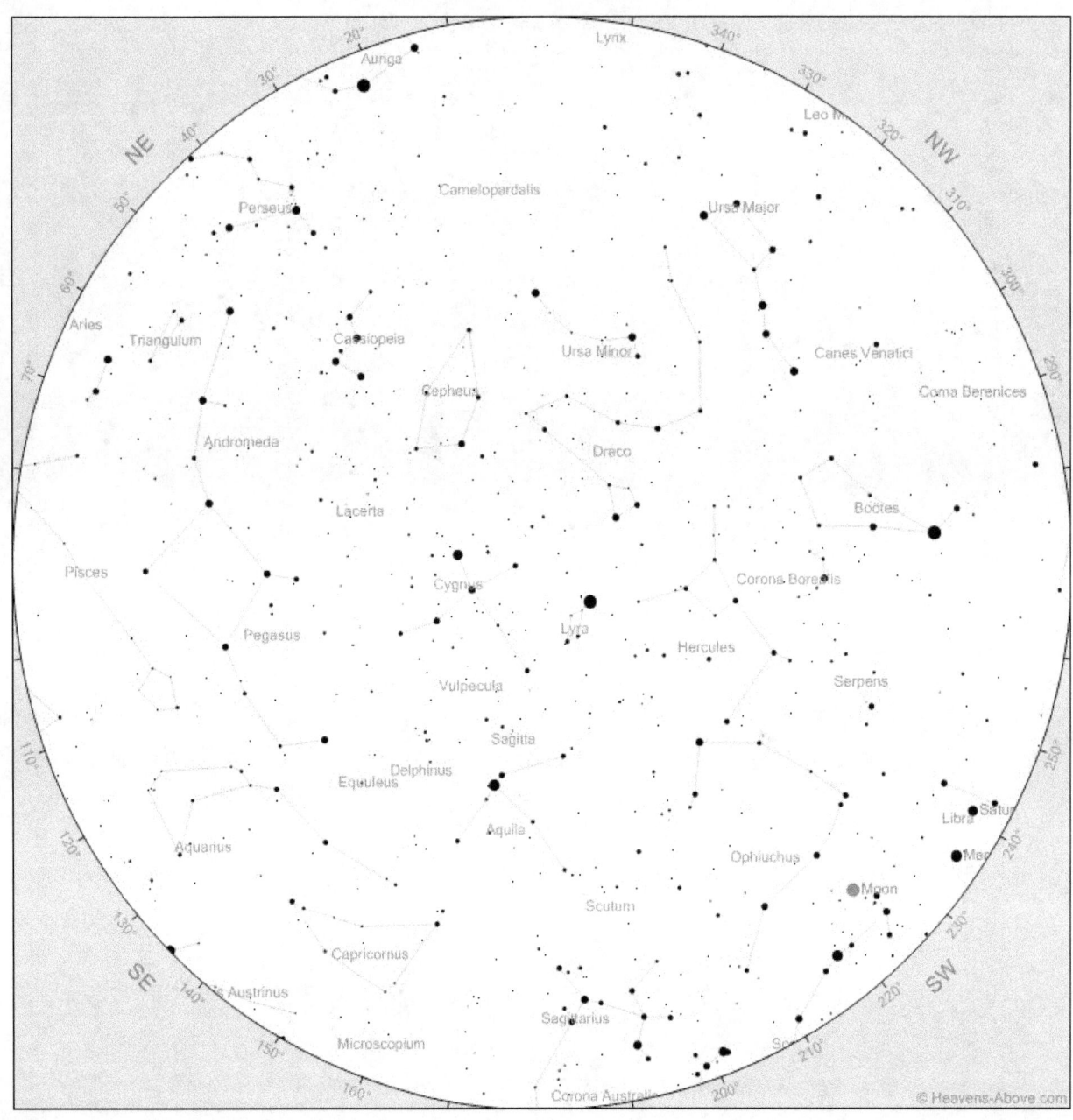

October 1st, 10 pm at 45°N

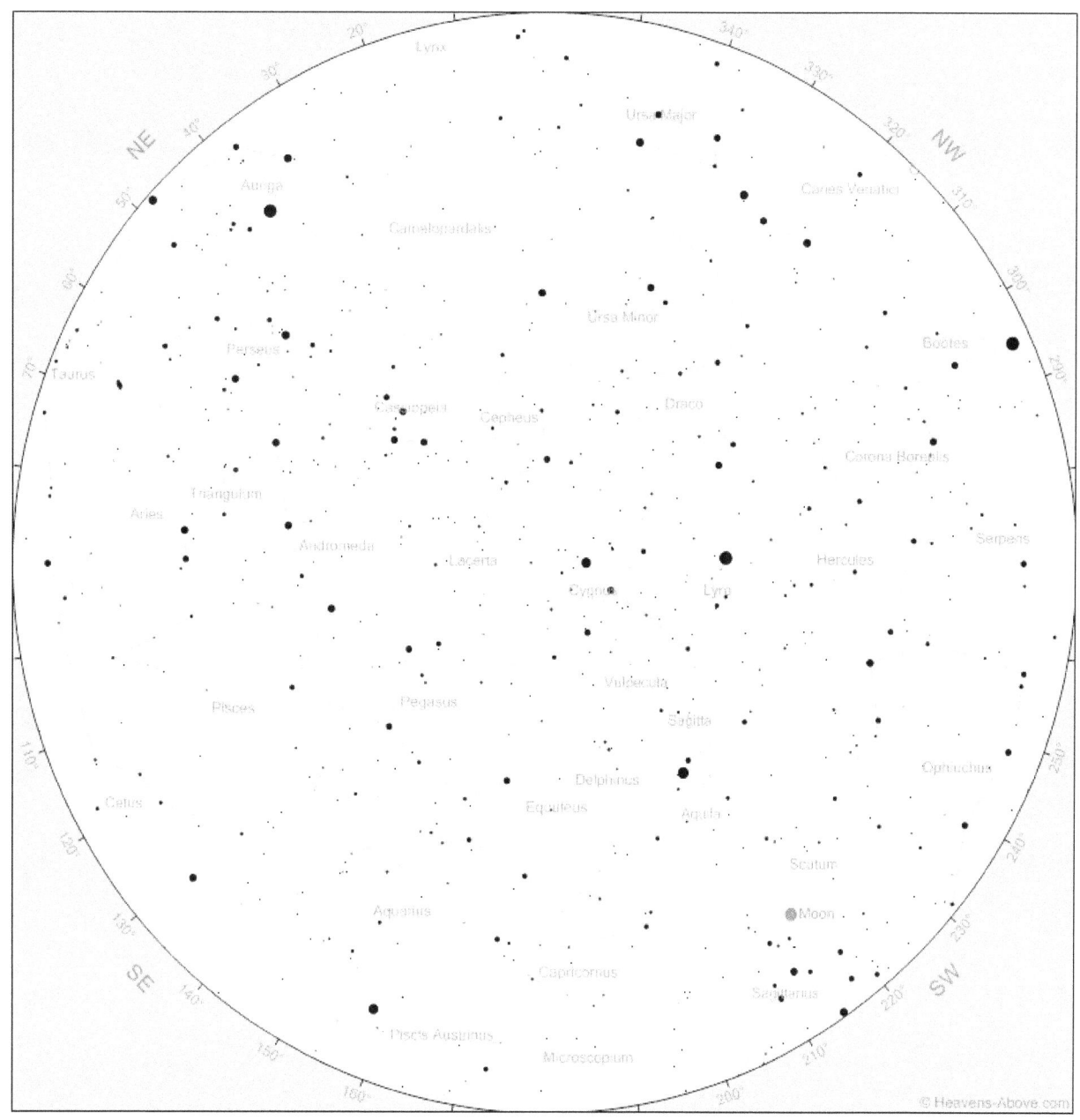

November 1st, 10 pm at 45°N

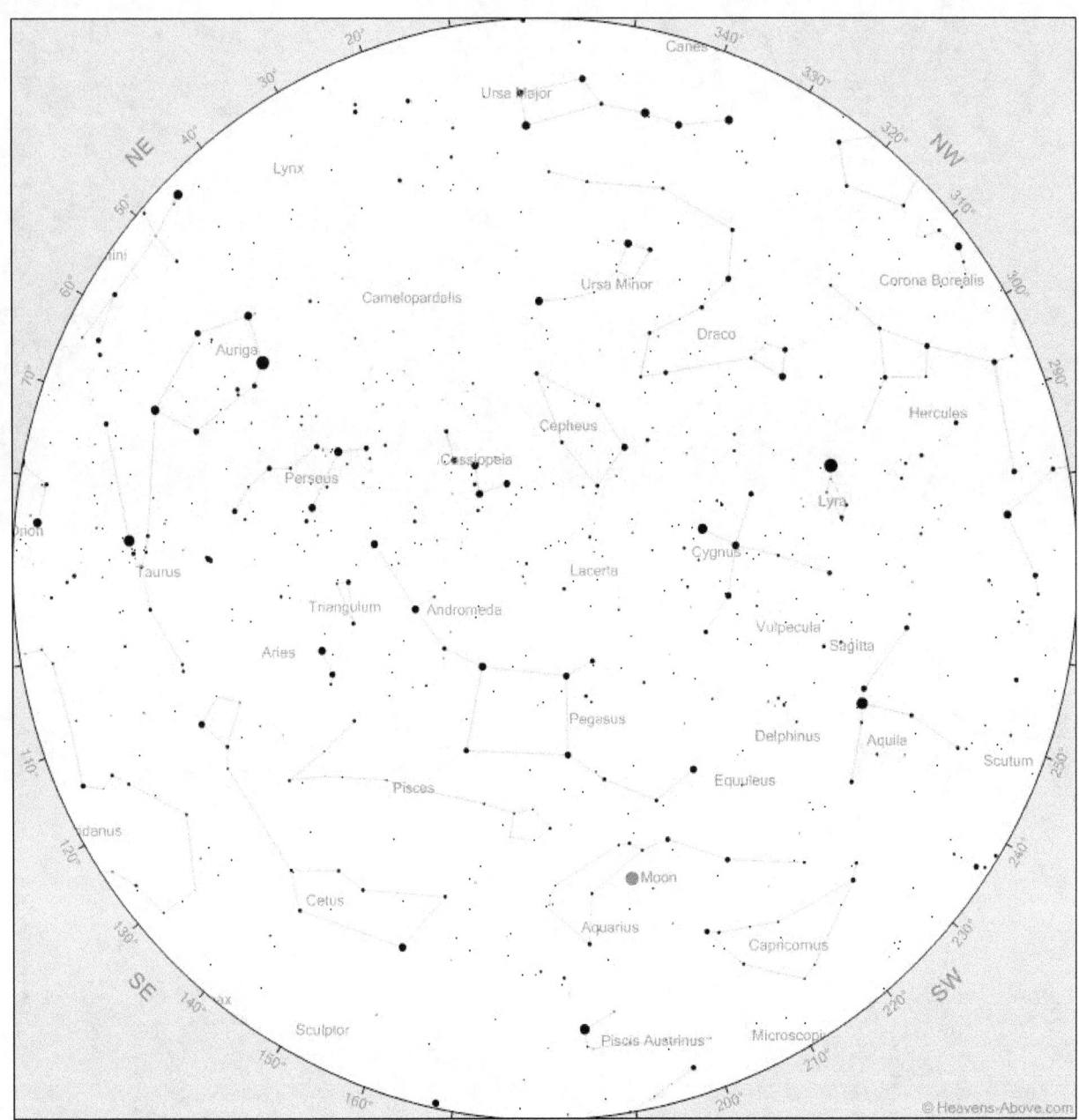

December 1st, 10 pm at 45°N

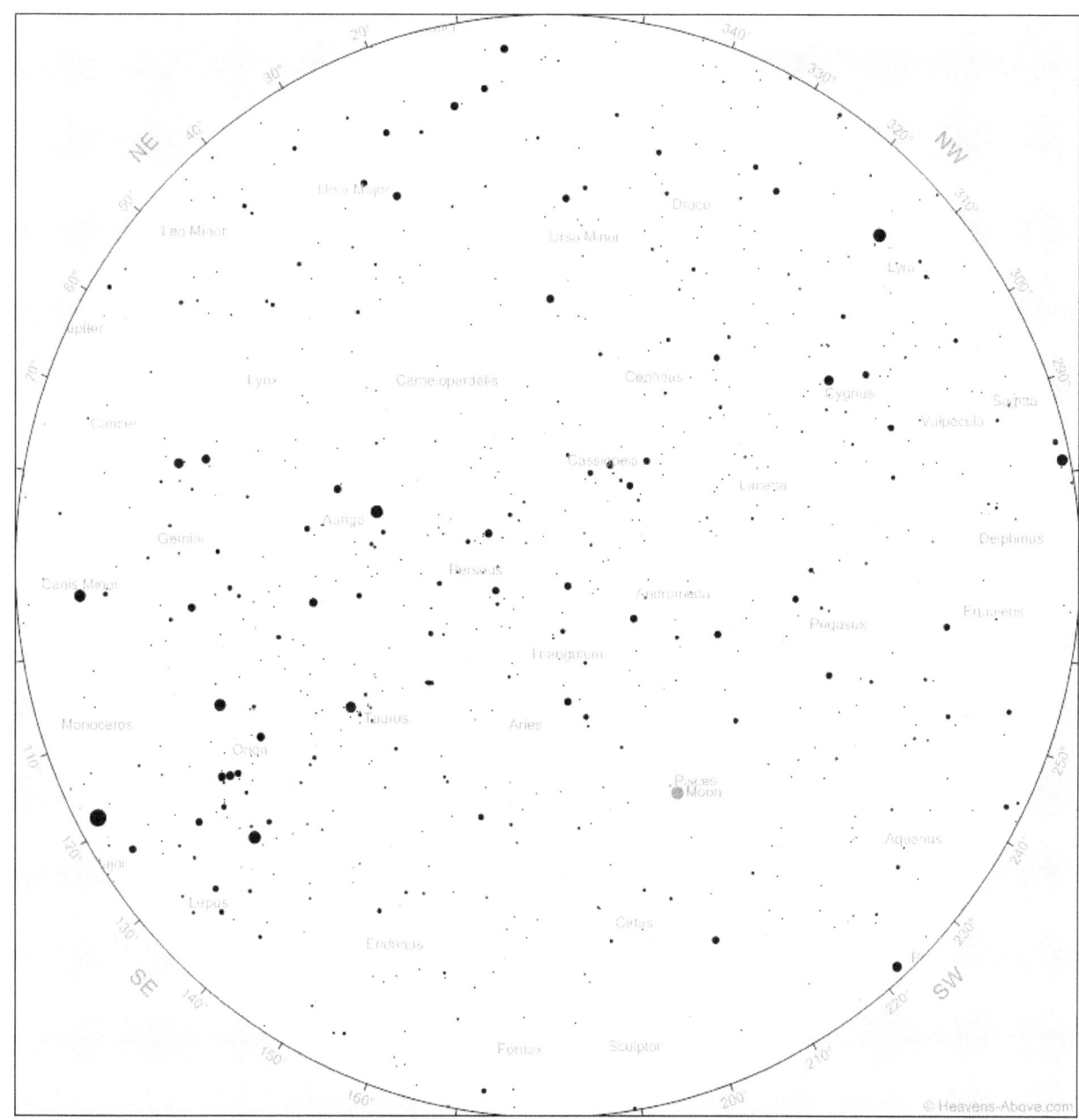

Activities

Designing an Observatory for Local Residents

Please print your name and sign next to it (only those present).

Leader (A)_____   _____

Explorer (B)_____   _____

Skeptic (C)_____   _____

Recorder (D)_____ _____

Your group has just received a $6000 grant for the purchase of observing equipment to help improve the exposure of city residents to astronomy. You promised in your grant application that you would be using your own means of transportation to take at least some of your equipment to local schools. You have an equipment room available at the local Museum for storage but will need to move the equipment outside for viewing. You must decide as to how you are going to spend the money and, because some of the money has been donated by Telescope.Com, you must make your selection from the accompanying catalogue supplied on the following pages. The grant requires you to list each item separately and give a detailed explanation for each purchase. (**Note**: see information on the back of this page.)

Item   **Justification**   **Cost**

                                                                                              _____
                                                                                              Total:

Amateur Observing Equipment

Type	Characteristics
Binoculars	- wide field of view—easy to find things - easy to use - excellent for getting acquainted with the sky - a useful accessory for use along with larger scopes - portable - anything over 8x requires a tripod
Newtonian reflector; Dobsonian mount	- uses a mirror for the objective - long tube contains the optics - excellent deep sky viewing (nebulae, galaxies, clusters, etc.) - simple mount does not track stars (no astrophotography) - more easily transported than equatorial mount - performance degraded in urban environment (light pollution)
Newtonian reflector; equatorial mount	- same optics as above but with more complicated equatorial mount that tracks stars and makes astrophotography possible - more expensive than Dobsonian
Apochromatic refractor; equatorial mount	- coated lenses produce unsurpassed lunar and planetary images - very expensive in larger sizes - best bet if you are limited to urban environments
Compound Scope; equatorial mount	- combination of lens and mirrors allows this to fit in a very short tube making it highly portable - wide range of accessories available including integrated computer drives - good all-round performer - images second only to refractor

Eyepieces: The eyepiece is what links the telescope's image to your eye; a poor quality eyepiece can severely limit the capabilities of an otherwise good telescope. Moreover, a selection of different eyepieces allows the magnification of the telescope to be adjusted (the useful range is about 40-200x for amateur scopes). A total of four eyepieces (average cost of about $70 each) would generally give an adequate selection of magnifications.

Solar Filters: One object in the sky that people sometimes forget about viewing is our Sun. A simple solar filter (about $100) can be added to a telescope to make it possible to safely view the Sun through the eyepiece. This is especially nice to have if you are talking to school groups during the day and want to give them a chance to see something through your telescope. One solar filter is needed for each different size (aperture) of telescope.

Binoculars: A good pair of binoculars costs about $100. A pair of 8x50's provides wide field views and can be hand-held. 15x60's do collect more light and allow you to see more detail but they require a tripod, which costs about $75.

Samples from Telescope.com

Refractors

Celestron PowerSeeker 60 AZ Telescope
★★★★
Sale Price: $59.98

Levenhuk Skyline 60x700 AZ Telescope
Our Price: $64.95

Levenhuk Strike 50 NG Refractor Telescope
Our Price: $74.95

Meade Infinity 102mm Altazimuth Refractor Telescope
Sale Price: $229.98

Celestron AstroMaster 90 AZ Telescope
★★★★
Sale Price: $229.98

Celestron Ambassador 50mm Brass Tabletop Telescope
Our Price: $249.95

Celestron Omni XLT 120 Refractor Telescope
★★★★
Our Price: $649.95

Celestron Advanced VX 6 Inch Refractor Telescope
★★★★★
Our Price: $1,299.00

Celestron Ambassador 50 Brass Telescope
Our Price: $399.95

Dobsonian Reflectors

Zhumell Z8 Deluxe Dobsonian Reflector Telescope

★★★★☆ 201 Reviews

Our Price: $399.00

Zhumell Z10 Deluxe Dobsonian Reflector Telescope

★★★★☆ 177 Reviews

Our Price: $549.00

Zhumell Z12 Deluxe Dobsonian Reflector Telescope

★★★★☆ 123 Reviews

Our Price: $699.00

Meade 12 Inch LightBridge f/5 Truss-Tube Dobsonain Telescope

☆☆☆☆☆ 0 Reviews

Our Price: $999.00

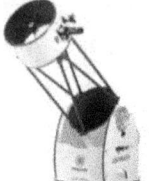

Meade 16 Inch LightBridge f/4.5 Truss-Tube Dobsonian Telescope

★★★★★ 1 Review

Our Price: $1,999.00

Sky-Watcher Traditional Dobsonian 10 Inch

☆☆☆☆☆ 0 Reviews

Sale Price: $585.00
List Price: $699.00

Sky-Watcher 14 Inch Dobsonian Telescope

★★★★★ 2 Reviews

Sale Price: $1,649.00
List Price: $1,899.00

Binoculars

ASTRONOMY BINOCULARS

Zhumell 20x80mm SuperGiant Astronomy Binoculars Package
★★★★★   570 Reviews
Our Price: $159.98

MID SIZE

Levenhuk Atom 8x40 Binoculars
★★★★☆   2 Reviews
Our Price: $34.95

GIANT

Celestron 15x70 SkyMaster Binoculars
★★★★★   749 Reviews
Sale Price: $69.98
List Price: $89.95

WATERPROOF ASTRONOMY BINOCULARS

Celestron Echelon 10x70 Binoculars
Sale Price: $629.98
List Price: $866.95

Cassegrain-Style Catadioptrics

Celestron NexStar 8SE Telescope
Ultimate Bundle

★★★★★ 134 Reviews
Sale Price: $1,299.98
List Price: $1,399.98

Celestron NexStar 4SE Telescope
Ultimate Bundle

★★★★★ 33 Reviews
Sale Price: $599.98
List Price: $649.98

Celestron NexStar 6SE Telescope
Ultimate Bundle

★★★★★ 102 Reviews
Sale Price: $899.98
List Price: $939.99

Celestron NexStar 8SE Telescope
with StarSense AutoAlign

★★★★★ 0 Reviews
Our Price: $1,528.95

Celestron 8 Inch CPC Schmidt-
Cassegrain Telescope Ultimate

★★★★★ 18 Reviews
Our Price: $2,099.99

Celestron Advanced VX 8 Inch
Schmidt-Cassegrain Telescope

★★★★★ 3 Reviews
Our Price: $1,599.00

Meade 10 Inch LX200-ACF f/10
Advanced Coma-Free Telescope

★★★★★ 0 Reviews
Our Price: $3,499.00

Meade 8 Inch LX200-ACF Advanced
Coma-Free Telescope

★★★★★ 2 Reviews
Our Price: $2,699.00

Celestron 11 Inch CPC Schmidt
Cassegrain Telescope

★★★★★ 34 Reviews
Our Price: $2,999.00

Meade 16 Inch LX600-ACF f/8
Telescope with StarLock MAX-

★★★★★ 0 Reviews
Our Price: $22,999.00

Meade 16 Inch LX200-ACF f/10
Advanced Coma-Free Telescope

★★★★★ 0 Reviews
Our Price: $15,999.00

Meade 16 Inch LX200-ACF f/10
Advanced Coma-Free Telescope

★★★★★ 0 Reviews
Sale Price: $13,499.00
List Price: $17,999.00

Eyepiece Sets

TELESCOPE EYEPIECES

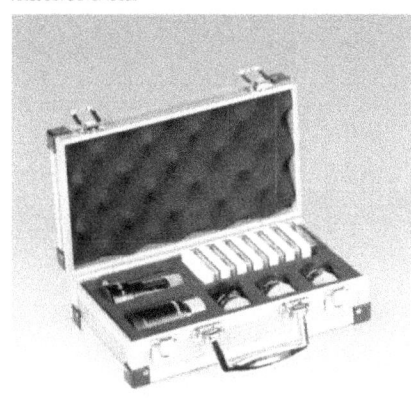

Zhumell Telescope 1.25 Inch Eyepiece and Filter Kit
★★★★⯪
Sale Price: $89.98

1.25 INCH EYEPIECES

Celestron Telescope Eyepiece and Filter Kit 1.25 Inch
★★★★★
Sale Price: $129.98

2 INCH EYEPIECES

Tele Vue Nagler Type 4 Telescope Eyepieces
★★★★★
Sale Price: $375.00

ZOOM EYEPIECES

Meade 8mm - 24mm Zoom Telescope 1.25 Inch Eyepiece
★★★★⯪
Sale Price: $69.98

SPECIALTY EYEPIECES

Meade View Plossl Illuminated Reticle 1.25 Inch Telescope Eyepiece
★★★★⯪
Sale Price: $69.98

EYEPIECE & FILTER KITS

Celestron Observers Accessory Kit
★★★★★
Sale Price: $39.98

1.25 INCH EYEPIECES

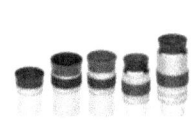

Celestron Omni Series Telescope Eyepieces
★★★★⯪
Sale Price: $19.98

EYEPIECES ON SALE

Zhumell Z Series Planetary Telescope Eyepieces
★★★★⯪
Our Price: $42.98

Meade Series 4000 Super Plossl Telescope Eyepieces
★★★★★
Our Price: $29.99

Celestron Telescope Eyepiece and Filter Kit 2 inch
★★★★⯪
Sale Price: $244.95

Vixen Lanthanum Wide LVW Telescope Eyepieces
★★★★⯪
Sale Price: $269.00

Baader Planetarium Hyperion Telescope Eyepieces
★★★★⯪
Our Price: $147.00

Solar Telescopes

iOptron Solar 60 Telescope with
Solar Filter and Electronic Eyepiece
★★★☆☆ 2 Reviews
Our Price: $399.00

Coronado SolarMax II 60 Double
Stack Solar Telescope with RichView
★★★☆☆ 1 Review
Our Price: $2,099.00

Coronado SolarMax II 90 Double
Stack Solar Telescope with RichView
☆☆☆☆☆ 0 Reviews
Our Price: $5,999.00

Sun Filter for Telescope

AstroZap Baader Solar Filters

★★★★★ 4.6 40 Ratings Read 40 Reviews Write a Review Customer Q&A

$90.00

Model Size

Meade 8 inch LXD75 SN

[1] **ADD TO CART**

This item is on backorder. It will ship in 15 days from date your order is placed.

- Solar filters made with Baader AstroSolar™ 18Ga film

Lenses and Telescopes

Please print your name and sign next to it (only those present).

Leader: (B)_____   _____

Explorer: (C)_____   _____

Skeptic: (D)_____   _____

Recorder: (A)_____   _____

Learning Objectives
1. Describe the devices that use lenses.
2. Determine how different lenses affect vision.
3. Measure the focal length of two lenses.
4. Construct a refracting telescope.

Part I: Exploring Lenses

1. List some inventions that use lenses?

2. How do lenses differ from each other?

3. Why do you think that some lenses magnify more than others?

Part II: Exploring Magnification

Have the group leader obtain two different lenses from the instructor. Using a pen and a small amount of masking tape, label the lens that is more curved—we say that the THICKER lens is the more CURVED lens—as LENS A and the lens that is less curved as LENS B. Place the tape-label on one side of the lens at the edge so that you can still look through the lens.

4. Hold lens A close to your eye to look at the writing on this page. Now use lens B. Which lens makes the letters appear larger (make sure the whole group agrees on this)? _____

5. Holding the lenses about 12 inches from your eye, look at the prints and patterns on your index finger using first lens A and then lens B. In the circles below, carefully make a detailed drawing of exactly what you see.

Person A	Person B	Person C	Person D
◯	◯	◯	◯
Lens A	Lens A	Lens A	Lens A
◯	◯	◯	◯
Lens B	Lens B	Lens B	Lens B

6. Hold lens A and lens B close to the letters on this page (less than 3 inches away). Does the writing appear right side up or upside down?

   A: _____

   B: _____

7. Stand up and leave the paper on the table. Hold one of the lenses close to your eye. Slowly move the lens closer and closer to the page until the letters become clear. Describe what you see. Repeat with the other lens. How do the lenses differ?

8. What do you think is the definition of *magnifying power*?

9. What do you think is meant by the term *field of view*? Does a lens that magnifies greatly have a large or small field of view compared to a lens that magnifies less?

10. Start with lens A about 4 inches from your eye and look at an object on the other side of the room. Slowly move the lenses **closer** and **farther** away from your eye. Are there any differences between what you see at close, medium, and long range?

11. Repeat step 10 using lens B.

A lens has a characteristic feature called a focal length. The focal length is the distance away from the lens that it will focus light. While sitting underneath a ceiling light in the room or against a wall opposite a window, adjust the distance of the lens from a piece of white paper until you can clearly see the image of the light source. Measure how far above the paper you must hold each lens to do this. This distance is called the focal length.

12. What is the focal length of lens A? _____

13. What is the focal length of lens B? _____

14. Did the lens with more magnification have the greater focal length? _____

15. What characteristic of a lens gives it a large magnification?

Part III: Using Your Telescope

16. Focal length of lens A + Focal length of lens B = Distance between lenses in a telescope.

    _____ + _____ = _____

Separate the lenses by a distance of the **sum of their focal lengths** with lens B nearer your eye. This kind of telescope is called a *refracting telescope*. Have each of your group members try the telescope.

17. A telescope uses two lenses. The lens that is closer to your eye is called the eyepiece lens and the lens that is closer to the object at which you are looking is called the objective lens. Note which lens is the objective lens and which is the eyepiece lens by drawing a detailed diagram of your telescope below (be sure that you label which lens is which).

18. Describe what you see with your telescope for both near and far objects. Can you alter the image by rotating or spinning the lenses?

19. Now switch the positions of the lenses A and B. How does it change what you see?

20. Which arrangement makes the more powerful telescope (i.e., higher magnification and smaller field of view)?

21. Which arrangement makes the telescope with the widest field of view?

22. When objects are viewed through your telescope, are they right-side-up or upside down?

23. How far away can you read this page with your telescope?

Star Charts

Please print your name and sign next to it (only those present).

Leader: (C)_____   _____

Explorer: (D)_____   _____

Skeptic: (A)_____   _____

Recorder: (B)_____   _____

Learning Objectives
1. Locate Polaris on a star chart.
2. Identify star names and asterisms on a star map.
3. Understand the nightly and seasonal motions of stars.

Your instructor will give you a star map for tonight's sky. Your task is to locate as many stars and asterisms as you can—your goal is find and label at least five of each. As a first step, use the star charts in Appendix G and begin by finding the North Star (Polaris). After you have completed your star map, answer the following questions.

1. How can you use the Big Dipper to find Polaris?

2. On a map of the United States, north is toward the top of the page and west is to the left. On all of the star charts, north is toward the top of the page and west is to the right. How do you account for this difference?

3. This star chart is for 10 p.m. tonight. By midnight tonight, the positions of the stars will have changed and will look like next month's chart at 10 p.m. Which chart should you use to see what tonight's sky will look like at 2 a.m.?

4. By comparing the appearance of the sky at 10 p.m. and 2 a.m., determine in which direction (east to west or west to east) a star near the **southern** horizon appears to move. Explain how you determined your answer.

5. By comparing the appearance of the sky at 10 p.m. and 2 a.m., determine in which direction (east to west or west to east) a star near the **northern** horizon appears to move. You will have to turn the chart upside down to see how the sky will appear. Explain how you determined your answer.

Trigonometric Parallax

Please print your name and sign next to it (only those present).

Leader: (D)_____   _____

Explorer: (A)_____   _____

Skeptic: (B)_____   _____

Recorder: (C)_____   _____

Learning Objectives
1. Apply the procedure of trigonometric parallax to measure distance.
2. Know how apparent motion depends on distance.
3. Understand why this is a difficult process for measuring stellar distances.
4. Understand how shifts in apparent position are related to distance.

Introduction: If you hold a pencil at arms length and alternately close your left and right eyes, the pencil appears to jump back and forth. This results from viewing the object from slightly different positions. Similarly, the positions of the stars should appear to shift as our viewing location is changed by the motion of the Earth around the Sun—we change location by about 186 million miles between January and July. This apparent shift in stellar positions is called stellar parallax. For more than a millenium, the lack of observation of stellar parallax provided the standard rebuttal to suggestions that the Earth might indeed orbit the Sun. Because the stars are so very far away, this parallax effect was not actually measured until 1837.

Equipment: ruler, protractor, and worksheets (final two pages of this activity)

Part I: The instructions below will show you how, on Worksheet 1, you can determine the distance to an object (a dot on the page) based only on measurements of angle. **You can do this without drawing lines or placing an instrument beyond the dotted line in the middle of the page!** Then you can make the measurement directly with a ruler and compare the accuracy of the two methods.

    Step 1: Draw a 20 cm line across the bottom of Worksheet 1 centered on the page and mark the ends A and B (an example is shown in Figure 1 on next page).
    Step 2: Mark a dot near the top center of the page to represent an object. Your task is to determine the distance from the midpoint of the line to the object.
    Step 3: Place your protractor at point A (left end of the line) making sure that the base of the protractor is along your 20 cm line. Measure the angular position of the object by sighting along the protractor. Record the angle you measure directly on Worksheet 1.
    Step 4: Repeat step 3 from position B (right side of the line) and again record your result.
    Step 5: On the back of Worksheet 1, draw a 2 cm line and mark the end positions A and B. This will be a scale diagram of the original situation scaled such that 1cm=10cm.
    Step 6: Use your protractor to draw a line through point A **at the angle you measured** in step 3 (refer to Figure 1).
    Step 7: Repeat step 6 from position B. You now have two lines drawn on your scale diagram and you know that the object is at the intersection. Circle the object's position on your scale diagram.

Step 8: Use your ruler to measure the distance from the **midpoint** of your 2 cm line to the object. Remembering that each 1 cm on your scale drawing represents 10 cm on the original, determine the actual distance to the object. This is your **calculated value** and was determined **indirectly** without accessing the object directly (i.e., you never had to go to the object). Record this on the front of Worksheet 1.

Step 9: Because you are measuring the distance to a dot on a page, as opposed to a star in the sky, you can double check your measurement directly using a ruler. On the original figure, measure the actual distance from the midpoint of the 20 cm line to the dot. This is your **measured value** and was determined **directly**. Record this on Worksheet 1.

Reflection: In the space below, compare your indirect and direct measurements and suggest reasons for any discrepancies.

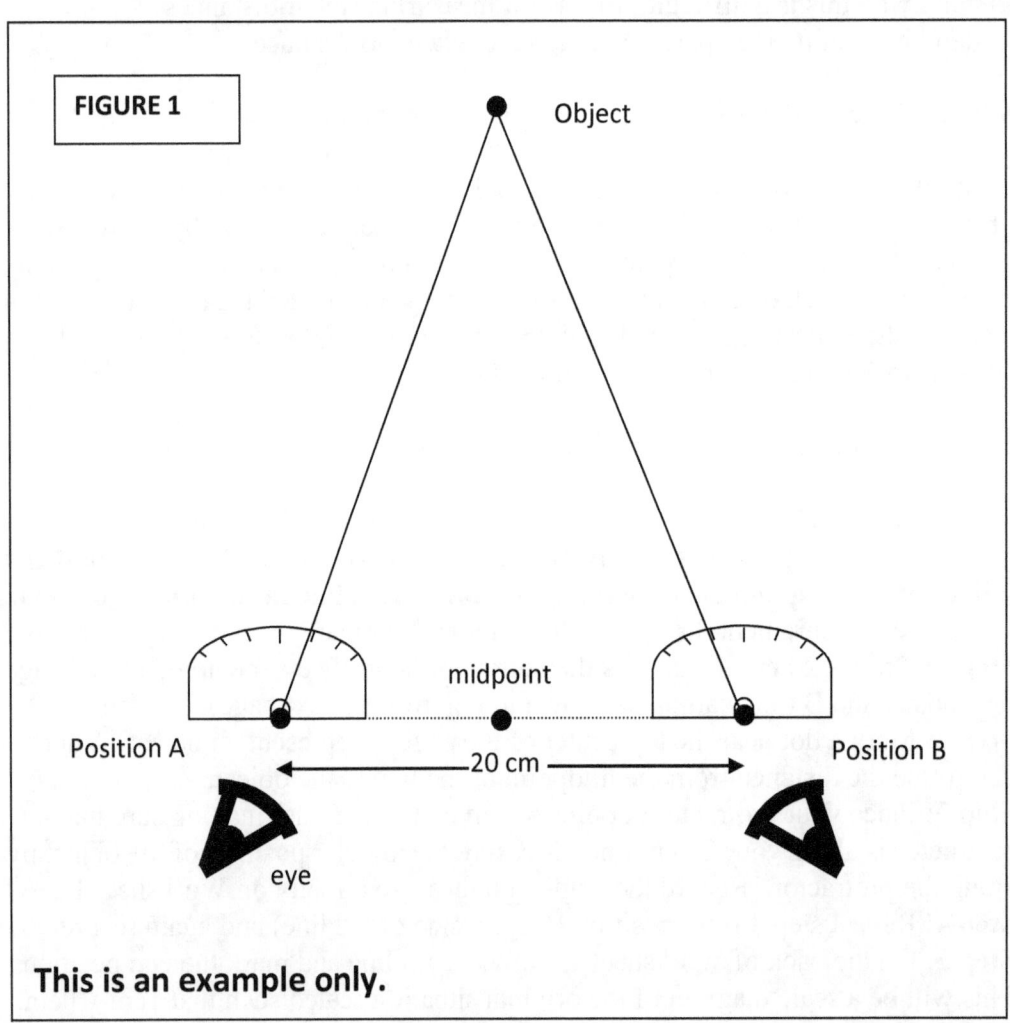

Part II: You are now going to repeat the same type of measurement but for an object too far away to directly measure easily—the same problem we have with stars. Use Worksheet 2 and place an object (the corner of a book or a pen) at a distance of 1 to 2 meters from the worksheet (the observer's location). As before, draw a 20 cm line at the bottom of Worksheet 2 and repeat the entire procedure from Part I to determine the distance to the object using angles only. Record all your work on Worksheet 2. Once you have used the parallax method (the indirect method using only angles) to determine the object's distance, measure the distance directly (this will not be easy, be creative) and record the result on the worksheet.

Questions:

1. How well did your parallax measurement (indirect) agree with the direct measurement?

2. Does the measurement become more or less difficult as the distance to the object increases? Explain.

Part III: The figure at right (not even close to scale) shows the Earth orbiting the Sun. There are two relatively close stars (A and B) as well as five distant stars. From Earth's location in January, Star A and Star B appear close to background star II. Six months later stars A and B appear next to different background stars.

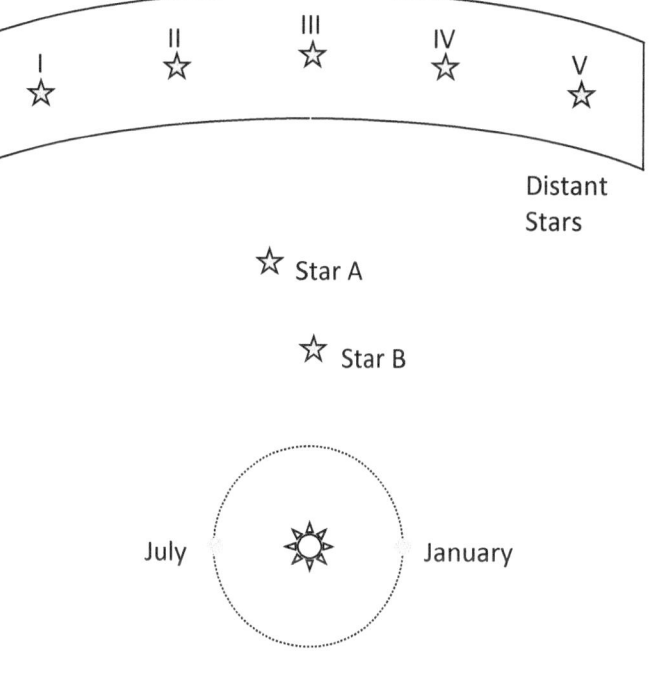

1. Next to which background stars would you look to see stars A and B in July? Draw lines on the figure to show how you determined your answer.

    A: _____

    B: _____

2. In general, which stars appear to move more relative to the really distant background stars, stars that are closer to Earth or stars that are slightly farther away?

155

3. The simulated photographs below represent show two pictures of the night sky taken six months apart. Which star is the most likely candidate to be relatively close? Explain, in detail, the rationale for your choice.

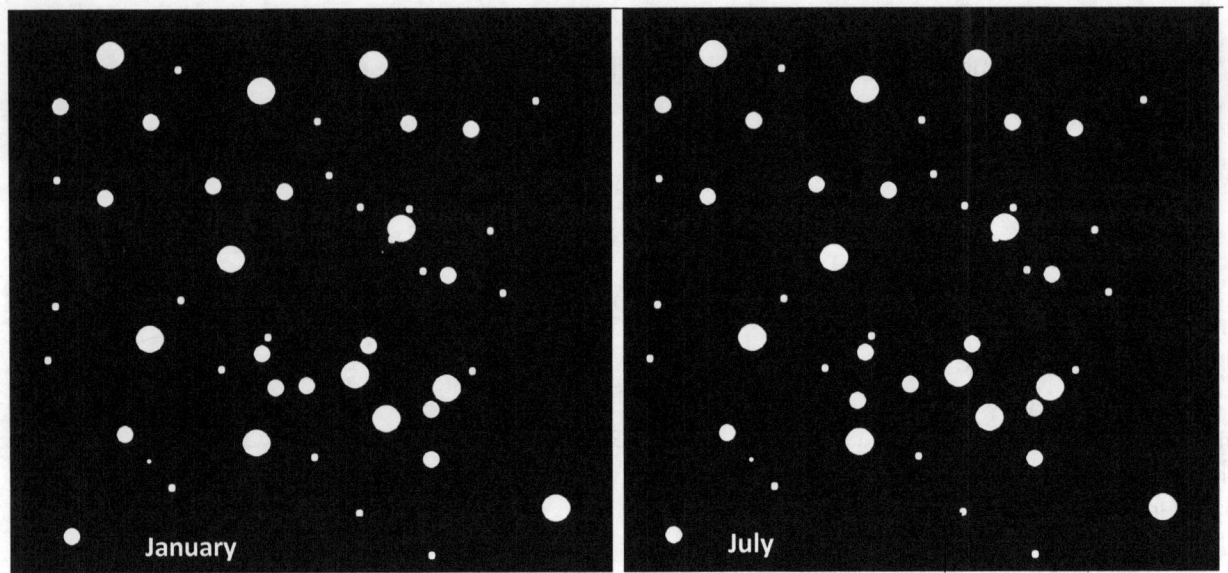

Worksheet 1

Angles to object
    position A:_____
    position B:_____

Distances
    calculated:_____
    (indirect)

    measured :_____
    (direct)

Draw a dot about here to represent your object.

Do not place anything above this line when performing the indirect measurement.

measure 20 cm along line, mark ends A and B

Worksheet 2

Angles to object
      position A:_____
      position B:_____

Distances
      calculated:_____
      (indirect)

      measured :_____
      (direct)

Stellar Spectra Classification

Please print your name and sign next to it (only those present).

Leader: (A)_____   _____

Explorer: (B)_____   _____

Skeptic: (C)_____   _____

Recorder: (D)_____   _____

Learning Objectives
1. Know that starlight is often broken up into component wavelengths with diffraction gratings to produce stellar spectra.
2. Understand how stellar spectra are classified as A, B, C, D, E and so on, based on prominent characteristics.
3. Understand how stellar spectra are related to composition and temperature.

Introduction: Classifying stars based on brightness is somewhat problematic. A star's apparent brightness can be affected by its distance from the observer, its size, or by the presence of interstellar dust. Instead, astronomers classify stars based on the major components of their spectra. Much like bar-codes on grocery store items, stellar spectra are each slightly different but have many characteristics in common. The current classification system used by astronomers was created by Annie Jump Cannon and Henrietta Leavitt working at Harvard in the early 20th Century. The study of spectra provides scientists with important information about stars that is otherwise inaccessible. This information includes composition and temperature.

Part I: Classifying Stellar Spectra

Included in this activity is a table of simulated stellar absorption spectra (page 169). Your first task is to sort the spectra by creating a classification scheme. As with real stellar spectra, you will never find two exactly the same. The thickness of each line represents how much light is removed at a particular wavelength, so the both the **thicknesses** and the **positions** of the lines are very important. Astronomers usually focus on the thickest lines first. Record your results in the table on the next page.

Note: The table contains five rows but you do not need to use them all. Also, there is no requirement that your classification scheme results in the same number of stars in each category.

Hint: *Imagine that the sixteen bar codes represent food items from four departments—meat, dairy, produce, and groceries. You might expect all the bar codes from the same department to look similar, but not identical. A similar task would be to sort them into groups representing the four departments.*

	Spectra ID Numbers	**Defining Characteristics** *(provide enough detail so that anyone could apply your scheme)*
Category I		
Category II		
Category III		
Category IV		
Category V		

Reflection: Find a nearby group and compare your classification scheme to theirs. What is the difference between how you chose to classify the spectra and how they did?

Part II: Matching Stellar Spectra

Now, let's try somebody else's scheme. Your task is to match the 16 unknown stellar spectra with the 4 known ones (standard spectra) at the end of this activity on page 172. You should identify four unknown spectra similar to standard A and the same for B, C, and D.

Known Spectra	A	B	C	D
Unknown Spectra Numbers				

Reflection: List below the specific characteristics of each of the four spectral families.

Standard A:

Standard B:

Standard C:

Standard D:

Part III: Determining Relative Stellar Temperatures

Blackbody radiation curves were acquired for all simulated stars, including the standards, and the results are listed on the Data Sheet on page 172. Remembering that the peak wavelength is a measure of the star's temperature, sort all twenty stars, including stars A, B, C, and D, into four new categories and record the star ID's in the table below. <u>The smaller the wavelength, the hotter the star.</u>

Hot Stars	Medium-Hot Stars	Medium-Cool Stars	Cool Stars

Reflection: Carefully compare and contrast the stellar temperature classification (table above) with the stellar spectra classification scheme used in Part II. How are the two classification schemes related?

Part IV: Determining the Temperature of a Star

At the bottom of the Data Sheet, you will find a table that relates the peak wavelength of the blackbody spectrum to the surface temperature of the star. Suppose a new star is discovered and its spectrum is shown below. Determine its temperature and justify your answer.

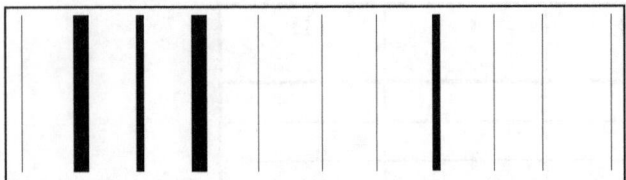

Part V: Self Check

Answer the following question first individually and then discuss it as a group. Check your answer with your instructor before leaving.

Q: What can we conclude about a star whose spectrum has a heavy concentration of lines at the right (long-wavelength) end?
  a. It is a particularly hot star.
  b. *It is a particularly cool star.*
  c. *There is no conclusion we can draw from this.*

SIMULATED STELLAR SPECTRA

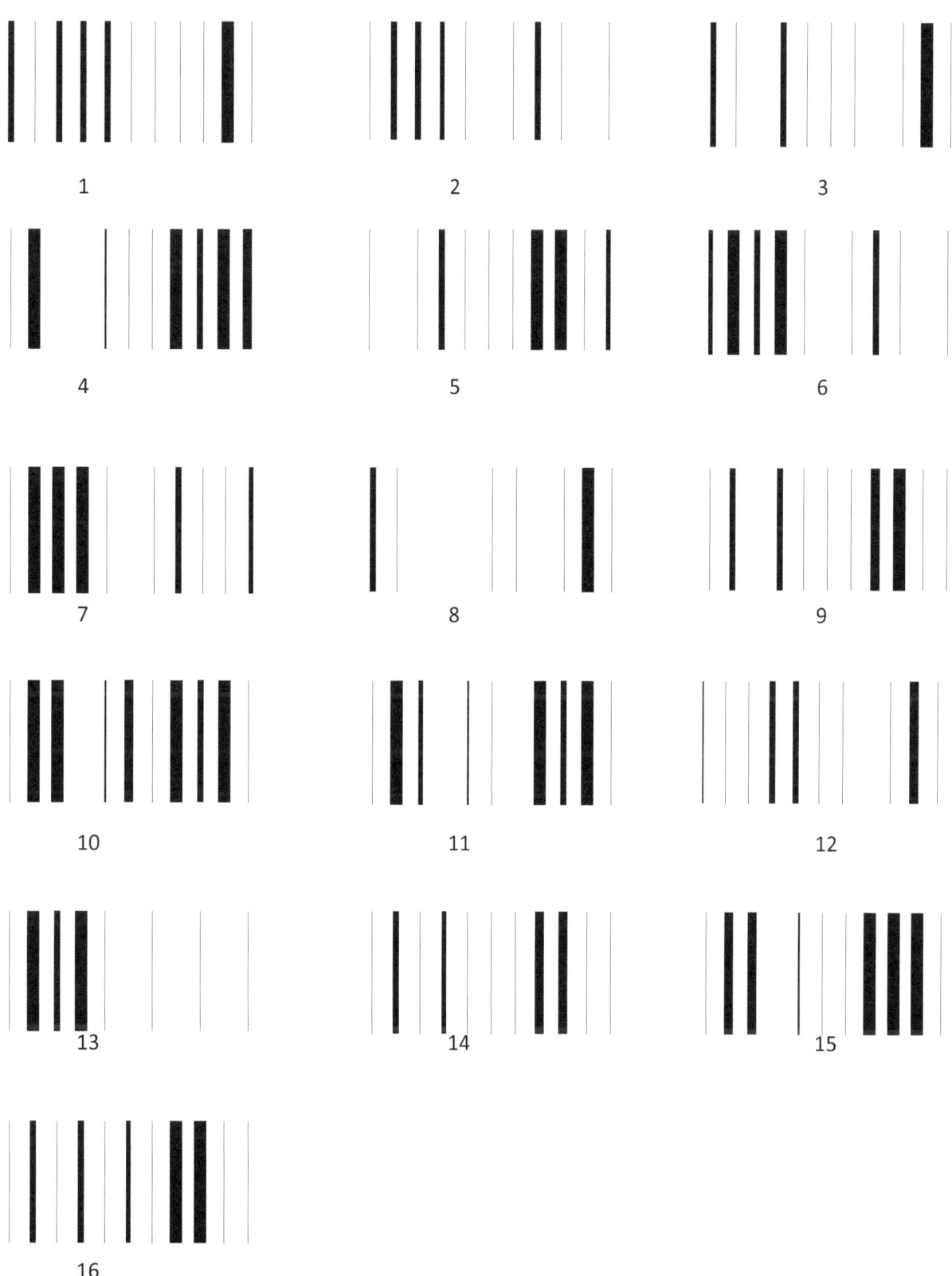

Stellar Spectra Classification Data Sheet

Standard Spectra Classifications

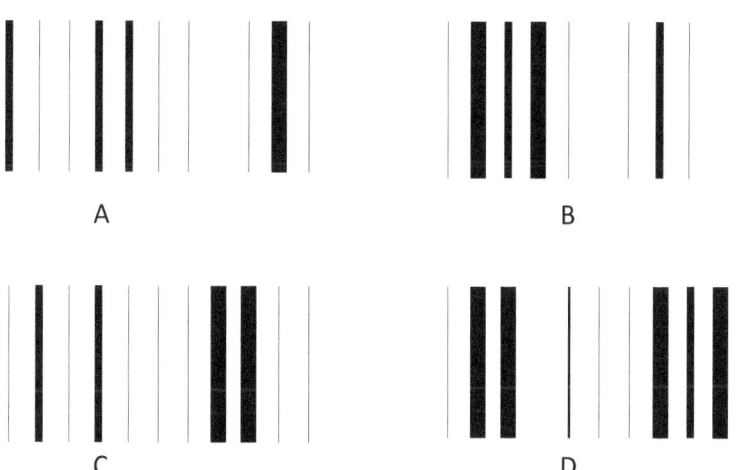

A   B

C   D

Blackbody Radiation Peak Values

Star ID	Peak Value (angstroms)
1	3625
2	2810
3	3612
4	7040
5	5470
6	2812
7	2790
8	3595
9	5510
10	6940

Star ID	Peak Value (angstroms)
11	7005
12	3610
13	2805
14	5515
15	7010
16	5555
A	3600
B	2800
C	5500
D	7000

Corresponding temperatures for four classes of stars studied.

Standard	Peak blackbody wavelength (angstroms)	Temperature (degrees)
A	3600	8000
B	2800	10,000
C	5500	5000
D	7000	4000

HR Diagrams

Please print your name and sign next to it (only those present).

Leader: (B)_____ _____

Explorer: (C)_____ _____

Skeptic: (D)_____ _____

Recorder: (A)_____ _____

Learning Objectives:
1. Know the parts of an HR diagram.
2. Know the names and properties of the major types of stars on the HR diagram.
3. Understand the scales and axes on the HR diagram.
4. Understand the relationship between luminosity, temperature, and size.
5. Understand how the properties of star types are determined.
6. Apply all of the above to describe a star.

Part I: Brief Review

1. Is apparent magnitude a measure of the actual amount of light being emitted by a star or just the amount we receive on Earth?

2. Is absolute magnitude a measure of the actual amount of light being emitted by a star or just the amount we receive on Earth?

3. Which is more closely related to a star's luminosity, the apparent magnitude or the absolute magnitude? Explain.

4. What is the relationship between a star's color and its temperature?

5. What does a star's spectral class tell you about the star?

Part II: Interpreting Correlations

The two graphs below, labeled A and B, represent two different data sets. One data set is the relationship between height and Intelligence Quotient (IQ) for a random collection of astronomers. The other is the relationship between height and body weight for the same group. Each diamond represents the data for one individual. Both data sets have been plotted with height on the horizontal axis with the other characteristic (weight or IQ) on the vertical axis.

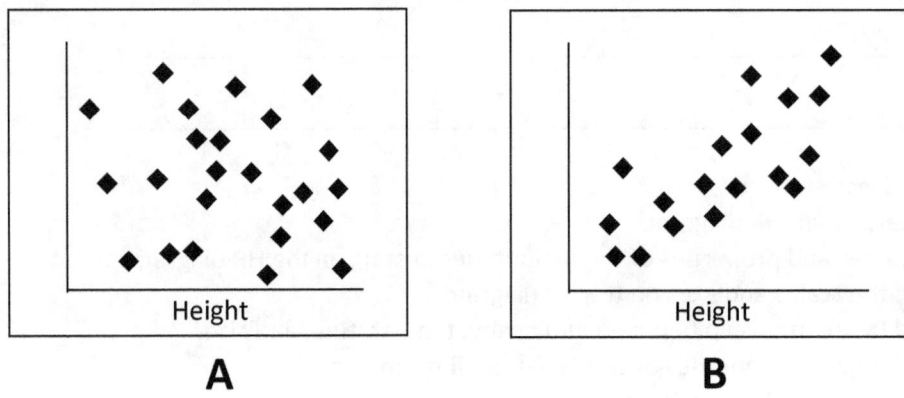

6. Identify which graph is which and explain your reasoning.

Part III: Brief Review of Spectral Classes:

Astronomers classify stars by their spectral type. If we consider the temperature of stars, as revealed to us by their color, we find that the O stars, are hottest, followed by B stars, A stars, F, G, K, and M stars. A star that has an intermediate temperature between an O star and a B star is called an O5 star. This temperature scheme is given from hottest to coolest as:

O0, O1, O2, O3, O4, O5, O6, O7, O8, O9, B0, B2, B3, B4, B5, B6,...

Our Sun is a G2 star, which means that it is an intermediate temperature star.

7. The star □-Cetus is a G8 star. Is this star hotter or cooler than our Sun?

Part III: Building the HR Diagram

Early this century, two researchers, Enjar Hertzsprung and Henry Norris Russell, independently developed what has come to be known as the Hertzsprung-Russell (HR) diagram. Tables I and II at the end of this activity list the spectral classes and luminosities for two sets of stars. The first set is 10 bright stars as seen from Earth and the second set is 10 stars near Earth.

The HR diagram is a plot of absolute magnitude/luminosity on the vertical axis versus spectral class/temperature/color on the horizontal axis for a collection of stars. Your task is to create an HR diagram from the data in Tables I and II. For each star, indicate its position on the graph below with a small "x." The Sun (G2, 4.8) has been plotted for you. Plot the bright stars (Table I) and near stars (Table II) in different colors by using, for instance, pencil for the bright ones and blue ink for the near ones. You do not need to label the stars. Each member of your learning group should plot about the same number of stars so that everyone gets some practice.

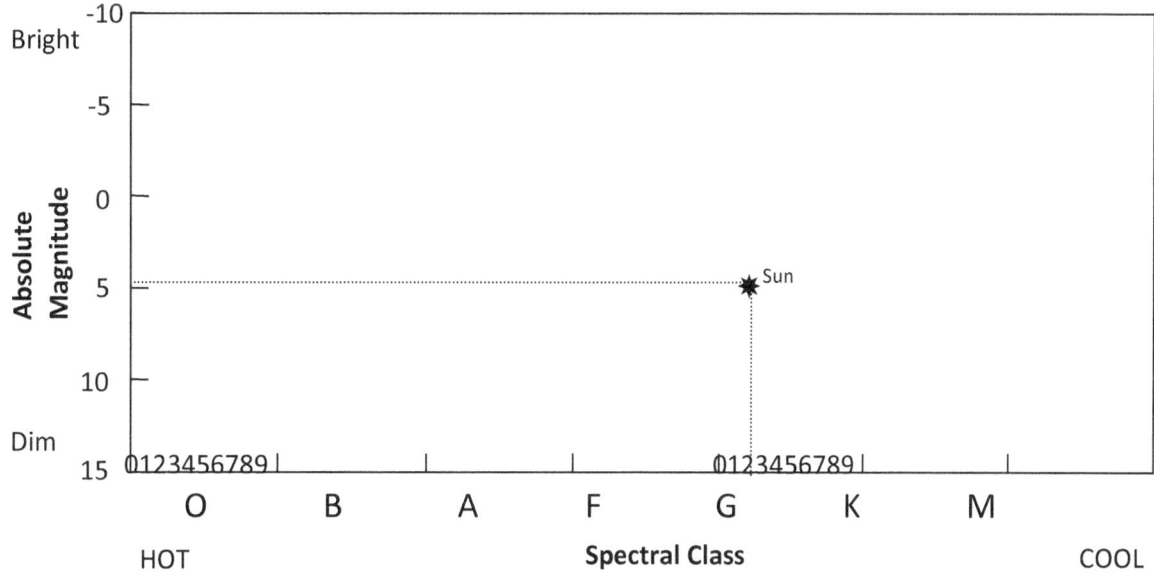

8. In general (i.e., you may have to ignore some of the data points), is there a correlation between spectral class and absolute magnitude? What is it?

9. Can you make any generalizations about stars that are close to Earth?

10. Most of the stars seem to be along a line from the upper left corner to the lower right corner of the HR Diagram. Stars that fall into this category of stars are called **main sequence stars**. Does our Sun fit into this category?

Part IV: Stellar Sizes

> **READ CAREFULLY:**
>
> Once we know a star's temperature and its total luminosity we can also deduce its size. The reason is that there is a connection between temperature and total energy output, which is described by the Stefan-Boltzmann law. The mathematical details are unimportant but there are two important aspects to remember:
>
> 1. The total amount of light energy that a star emits—**called the luminosity and measured by the absolute magnitude**—increases with temperature. In fact, it increases as the fourth power of the temperature so that a star the same size as our Sun but twice the temperature would emit 16 times the energy!
>
> 2. The amount of energy that a star emits per acre of surface area depends only on the star's temperature. Therefore, the total luminosity of a star increases with the number of surface acres. A star at the same temperature as our Sun but four times the surface area emits four times the energy.
>
> To estimate the size of a star we first determine its temperature from either its color or spectral class. This tells us how much energy each acre of the surface is emitting. The total luminosity is just a measure of the total energy, which allows the number of surface acres to be determined, which is a measure of the star's size.

11. Consider the stars in Tables I and II.

    a. Does Sirius B have a higher, equal, or lower surface temperature than Rigel? Explain your reasoning. (*hint: compare their spectral classes*)

    b. Compare absolute magnitudes to determine which of these two stars has the greater luminosity. Explain your reasoning.

    c. Which is the larger of the two stars? Explain your reasoning.

Part V: Classifications on the HR Diagram.

12. How does the size of stars near the top left of the HR diagram compare with stars of the **same luminosity** near the top right of the HR diagram? Explain your reasoning.

13. How does the size of stars in the top left of the HR diagram compare with stars at the **same temperature** near the bottom left of the HR diagram? Explain your reasoning.

14. Classify the following newly discovered stars as main sequence, dwarfs, or giants.

    a. Hermanson A (G4 +5.2):

    b. Slataurus (K8 –4.0):

    c. Adamisus (A3 +10.4):

    d. Franciscus G (F2 +3.1):

Table I: Bright Stars (as they appear from Earth)

NAME	SPECTRAL CLASS	ABSOLUTE MAGNITUDE
Sirius A	A1	1.4
Canopus	F0	-3.1
Arcturus	K2	-0.3
Capella	G3	-0.7
Rigel	B8	-6.8
Betelgeuse	M2	-5.5
Antares	M2	-4.5
Spica	B1	-3.6
Deneb	A2	-6.9
Procyon A	F5	2.6

Table II: Near Stars

NAME	SPECTRAL CLASS	ABSOLUTE MAGNITUDE
Sun	G2	4.8
Alpha Centauri A	G2	4.4
Sirius B	B8	11.6
Ross 154	M5	13.3
Ross 248	M6	14.8
Luyten 789-6 A	M6	14.6
Ross 128	M5	13.5
61 Cygnus A	K5	7.6
61 Cygnus B	K7	8.4
Procyon B	A0	13.0

176

Spectroscopic Parallax

Please print your name and sign next to it (only those present).

Leader: (C)_____   _____

Explorer: (D)_____   _____

Skeptic: (A)_____   _____

Recorder: (B)_____   _____

Learning Objectives
1. Know how apparent and absolute magnitudes are used to determine distance.
2. Use the HR diagram to determine the distance to stars.

Part I: Review

Betelgeuse has an apparent magnitude of +0.4, which tells us how bright it appears at its true location. Betelgeuse has an absolute magnitude of –5.5, which tells us how bright it would appear if we could move it to a distance of 10 parsecs (about 33 light-years).

1. Where would Betelgeuse appear brighter, in its true location or if were at a distance of 10 parsecs? Explain your reasoning.

2. So, is its true location closer or farther than 10 parsecs? Explain your reasoning.

Part II: Determining Relative Distances

Below is a list of five stars and their apparent and absolute magnitudes. Your task is to classify each star as being located either closer or farther than 10 parsecs.

	Apparent Magnitude	Absolute Magnitude	Distance (closer or farther)
Sirius	-1.5	+1.4	
Rigel	+0.14	-6.8	
Procyon	+0.37	+2.6	
Betelgeuse	+0.41	-5.5	
α Centauri	-0.01	+4.4	

Part III: Spectroscopic Parallax

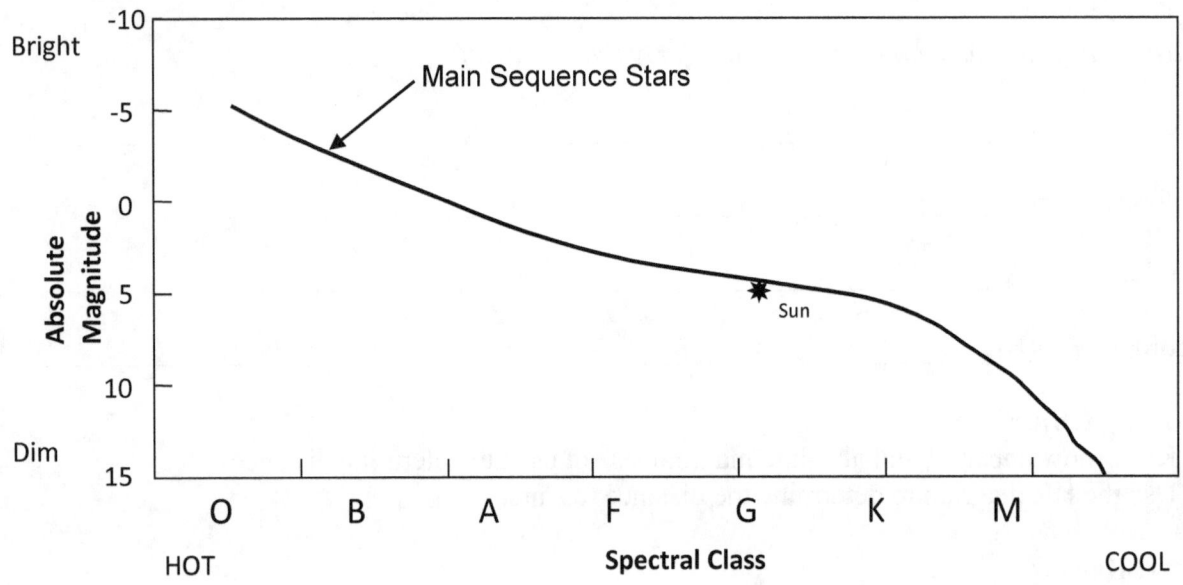

Below, is a table giving both the apparent magnitude and spectral class for five **main sequence** stars. For each star, do the following:
1. Using the above HR diagram, estimate the absolute magnitude for each star.
2. Complete the table by classifying each star as being **closer**, **slightly farther**, or **much farther** than 10 parsecs away. This procedure, called spectroscopic parallax, provides astronomers with a new strategy to measure the distance to stars.

Star	Apparent Magnitude	Spectral Class	Absolute Magnitude	Distance Estimate
Rigel Kentaurus	0.0	G2		
Vega	+0.04	A0		
Rigel B	+6.6	B9		
Canopus	-0.72	F0		
Deneb	+1.26	A2		

Note: In completing this table, what you have done is get an estimate of the distance to a star by comparing the apparent and absolute magnitudes without using a formula. The exact distance could actually be calculated using the formula $d=10^{(m-M+5)/5}$ pc, where m is the apparent magnitude and M is the absolute magnitude.

Reflection: A friend of yours who has never taken an astronomy class asks you how astronomers know the distances to stars. Explain how distances can be determined using spectroscopic parallax (i.e., the measurements that are made and how are they used).

Stellar Evolution

Please print your name and sign next to it (only those present).

Leader: (D)_____ _____

Explorer: (A)_____ _____

Skeptic: (B)_____ _____

Recorder: (C)_____ _____

Learning Objectives
1. Describe what determines the evolution rate of stars.
2. Know the evolutionary sequence of low and high mass stars.
3. Develop a concept map of stellar evolution.
4. Differentiate between novae, supernovae type I, and supernovae type II.

Concept maps are a strategy to create logical structures representing complex ideas. Based on a particular theme, a concept map uses bubbles and connecting lines to show connections between concepts. Consider the following seemingly simple example:

> *Birthday parties are celebrations given every year to commemorate the date of an individual's birth. Frequently, birthday celebrations begin with group singing of an age old and copyrighted song appropriately called, "Happy Birthday." Invited guests typically bring unsolicited gifts wrapped in colorful paper with unknown contents. Parties often conclude with desserts, such as cake and ice cream, and games of chance. Interestingly, birthday parties vary in structure, length, and atmosphere depending on the age of the recipient, typically becoming less structured in later years, with the exception of ten-year increments after the initial twenty-one birthday celebrations.*

Alternatively, a concept map might represent the above idea in the following way:

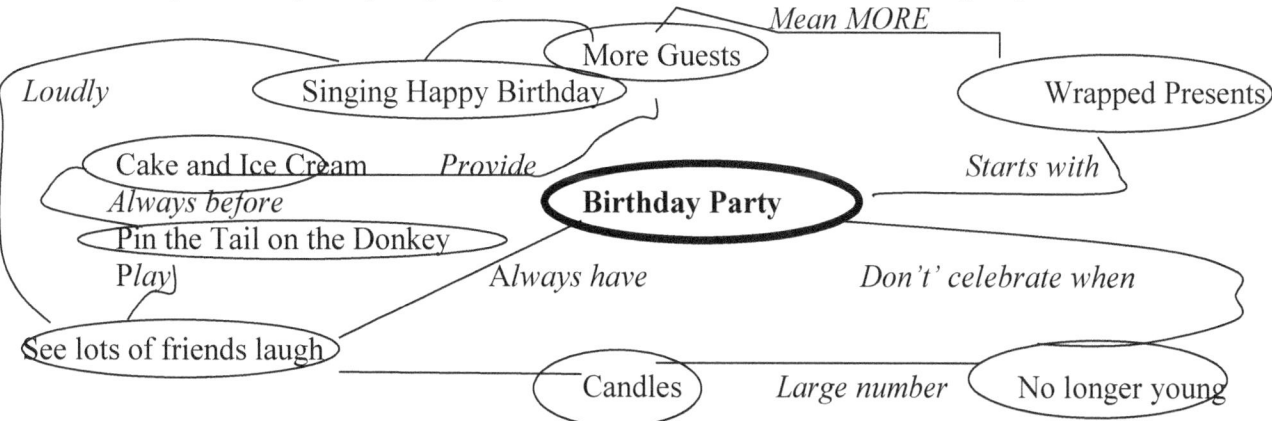

Your group task is to create, on the reverse side, a complex concept map for stellar evolution. Be certain to clearly label both **bubbles** and **lines**. The strategy to use is for <u>each person in the group to take turns and fill in one bubble and label a connecting line</u>. It is not necessary, or even preferable, to always connect the most recently created bubble. At the conclusion, be certain that all of the main ideas of stellar evolution are included.

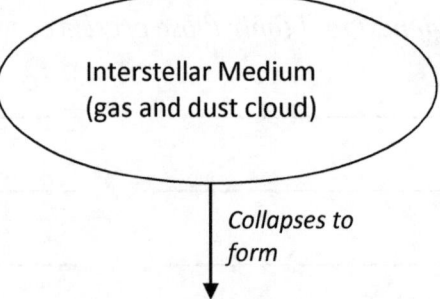

Galaxy Classification

Please print your name and sign next to it (only those present).

Leader: (A)_____   _____

Explorer: (B)_____   _____

Skeptic: (C)_____   _____

Recorder: (D)_____   _____

Learning Objectives
4. Know that galaxies are collections of billions of stars.
5. Understand that galaxies take a variety of forms.
6. Understand that galaxies are classified according to their appearance in four major categories.

Introduction: When stars are viewed through a telescope, they continue to appear as bright points of light without any apparent size or structure. However, there are some objects in the sky that, viewed through a telescope, look like "fuzzy" clouds. Some of these are the star forming regions called nebulae, which we have already discussed. Others, like those shown in the Hubble Space Telescope image to the right, are actually islands of stars that are much farther from us than the individual stars we see in the night sky. Although Immanual Kant first advanced the idea of "island universes" during the eighteenth century to explain observed compact clouds, it was not until the early twentieth century that astronomers began to develop an understanding of the nature of galaxies.

Image courtesy of NASA.

Part I: Developing a Classification Scheme

Included with this activity are eighteen galaxy photographs. Your first task is to sort the galaxies by creating and applying a classification scheme based on appearance. Use the table on the following page to record your results. The table has space for four categories but there is no requirement that you use all four. It is important to remember that, by focusing on different properties of the galaxies, it is expected that different groups will develop widely different classification schemes. There are no right answers at this stage of the scientific process although ultimately some classification schemes are found to be more productive than others.

	Galaxy ID Numbers	**Defining Characteristics** *(provide enough detail so that anyone could apply your scheme)*
Category I		
Category II		
Category III		
Category IV		

Part II: Applying Hubble's Classification Scheme

After you have completed Table 1, refer to Box 7-1 in text section 7.1 for Edwin Hubble's classification scheme, which was developed in the 1920s. Complete the following table according to his scheme.

Hubble's Categories	**Galaxy ID Numbers**	**Defining Characteristics** *(describe the characteristics used by Hubble)*

Part III: Questions

1. Unless there is an underlying model, classification systems are completely arbitrary as long as the defining characteristics are clear to everyone. Which system, yours or Hubble's, would be easier to explain to another learning group? Why?

2. Hubble viewed the tuning fork diagram as representing an evolutionary sequence for galaxies— all galaxies started as E0 and evolved to the right with time. How would Hubble have arranged the following three galaxies in order from youngest to oldest? Explain your reasoning.

Youngest: ___

Middle: ___

Oldest: ___

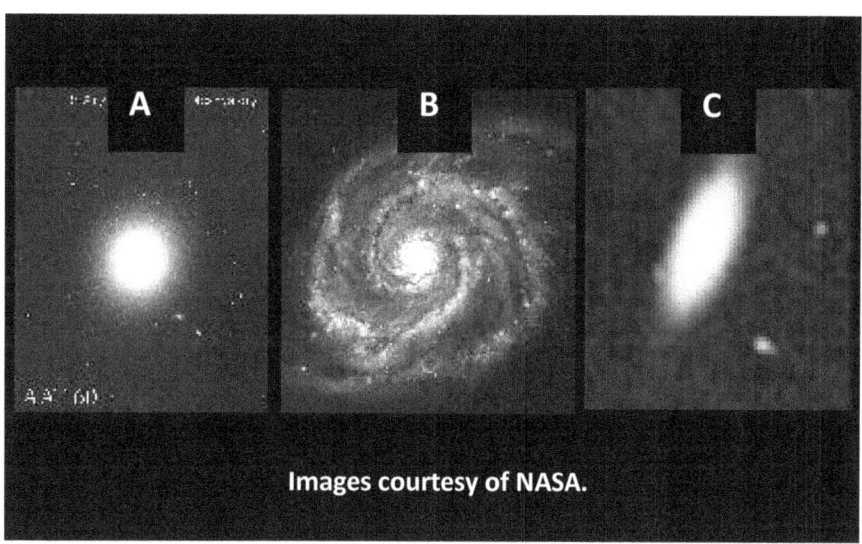

Rationale:

3. Astronomers now realize that the tuning fork diagram does not represent an evolutionary sequence. Does this mean that Hubble's scheme is useless? Explain.

GALAXY CHART
Page 1

GALAXY CHART
Page 2

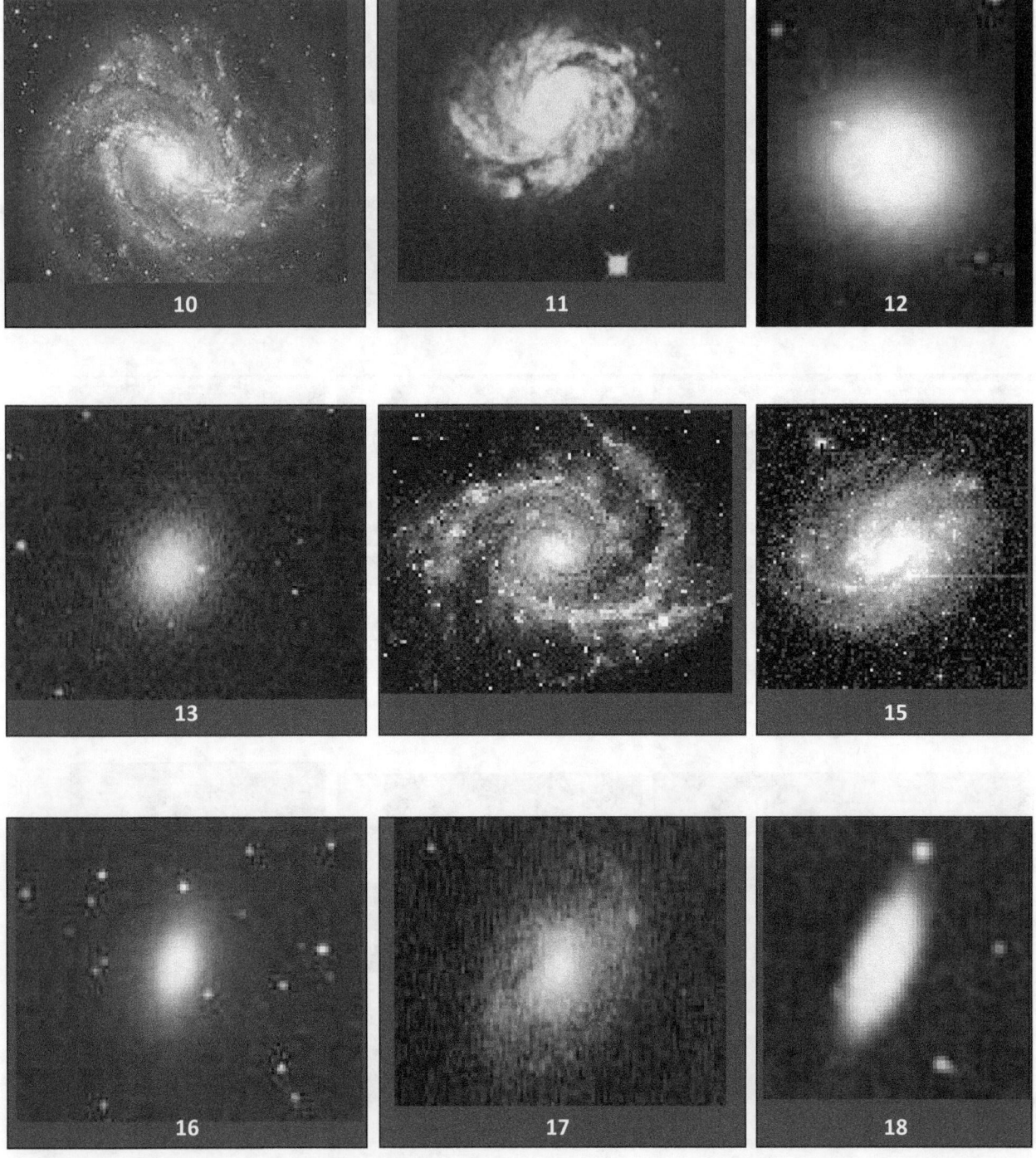

All galaxy images courtesy of NASA.

Hubble's Law

Please print your name and sign next to it (only those present).

Leader: (B) _____ _____

Explorer: (C) _____ _____

Skeptic: (D) _____ _____

Recorder: (A) _____ _____

Learning Objectives
7. Understand how Cepheid variables are used to estimate galactic distances from the period-luminosity curve.
8. Know how the recessional velocities of galaxies are determined.
9. Interpret the velocity versus distance plot for galaxies.

Part I: Cepheid Variables

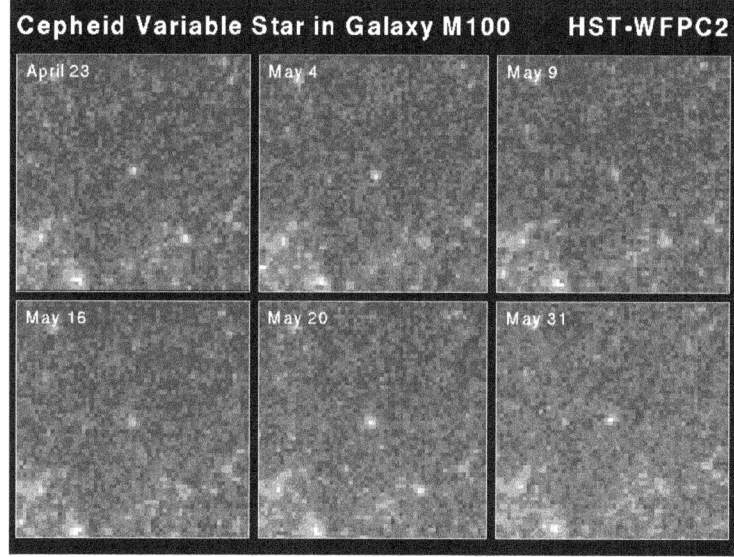

The photograph at right is a NASA Hubble Space Telescope image of a variable star in galaxy M100 (Courtesy of NASA). You can see that the brightness of the star changes with time (see the date on each image). These variable stars help astronomers calculate distances to other galaxies. Before 1910, Harvard astronomer Henrietta Leavitt began measuring the brightness of stars in a class known as Cepheid variables—bright stars with masses 5 to 20 times that of our Sun. Variable stars are so named because their luminosities, and therefore their apparent magnitudes, vary with time. Cepheid variables—or simply, Cepheids—are periodic variables that cycle through a complete bright-dim-bright cycle in times ranging from days to months, although they are typically in the range of weeks. What makes Cepheids so important to astronomers is that there turns out to be a direct relationship between a Cepheid's **period** and its **average luminosity**—the longer the period, the brighter the star. The figure on the next page presents data for twelve <u>different</u> Cepheids. Although there is some scatter in the data, it is clear that knowledge of the period does allow one to predict the luminosity.

1. Use the graph at right to determine which star takes longer to go through one complete bright-dim-bright cycle, Star A or Star B.
   Answer: _____

2. Which of the two stars, A or B, has the greater number for its average absolute magnitude? Explain your answer.

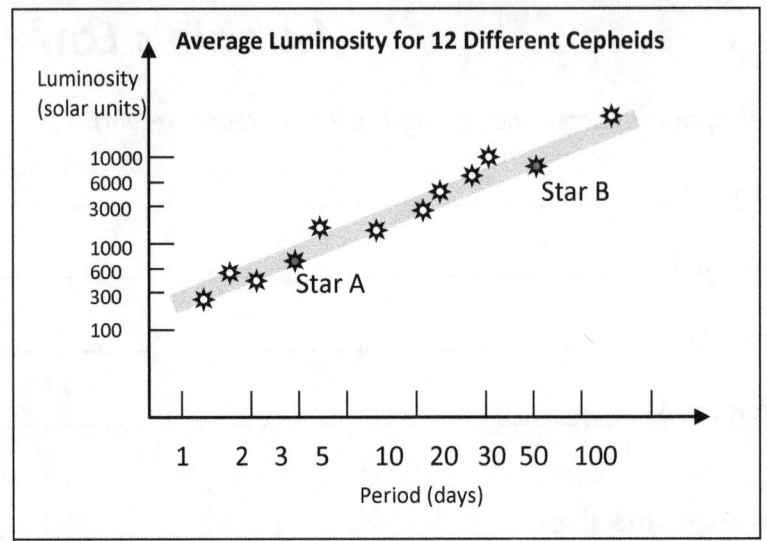

3. Use the series of HST images of the Cepheid variable in the galaxy M100 (previous page) to estimate the period of the star (the period is the number of days for one bright-dim-bright cycle).
   Answer: _____

4. From the Period-Luminosity graph (above), estimate the luminosity of the star.
   Answer: _____

5. Imagine another galaxy—call it SBGC98—contains a Cepheid with the same period as the one in M100. If the Cepheid in SBGC98 appears brighter in the sky than the Cepheid in M100, which galaxy is closer? Explain your reasoning.

6. In answering Problem 3 you effectively estimated the distance to the galaxy SBGC98 (except for doing some math) using a Cepheid variable as a standard candle. This process only requires two measurements to be made on a specific Cepheid to find its distance. What are these two measurements?

Part II: Red Shift

Another property of galaxies—significantly easier to measure than distance—is recessional velocity (how fast it is moving away from us). When a light source is receding, the observed wavelengths of standard emission and absorption lines

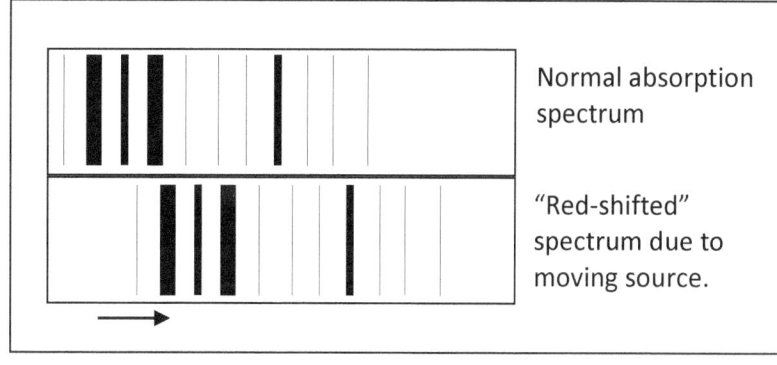

Normal absorption spectrum

"Red-shifted" spectrum due to moving source.

no longer appear where you would expect. The fact that the source is moving away causes the positions of the lines to shift to longer wavelengths, an effect known as red shifting. This phenomenon, similar to the changing pitch of a passing siren, is called the Doppler effect.

The table at right gives the distances and recessional velocities for five different galaxies. The galaxies are actually members of galactic clusters whose names are taken from the constellations

CLUSTER GALAXY IN	DISTANCE IN Mpc (millions of parsecs)	RECESSIONAL VELOCITY (km/s)
Virgo	19	1210
Hydra	1200	61,200
Bootes	770	39,300
Ursa Major	300	15,000
Corona Borealis	430	21,600

in which they appear. Indicate the position of each galaxy on the graph below with a small **X.** Then, using a ruler, draw a straight line through the data.

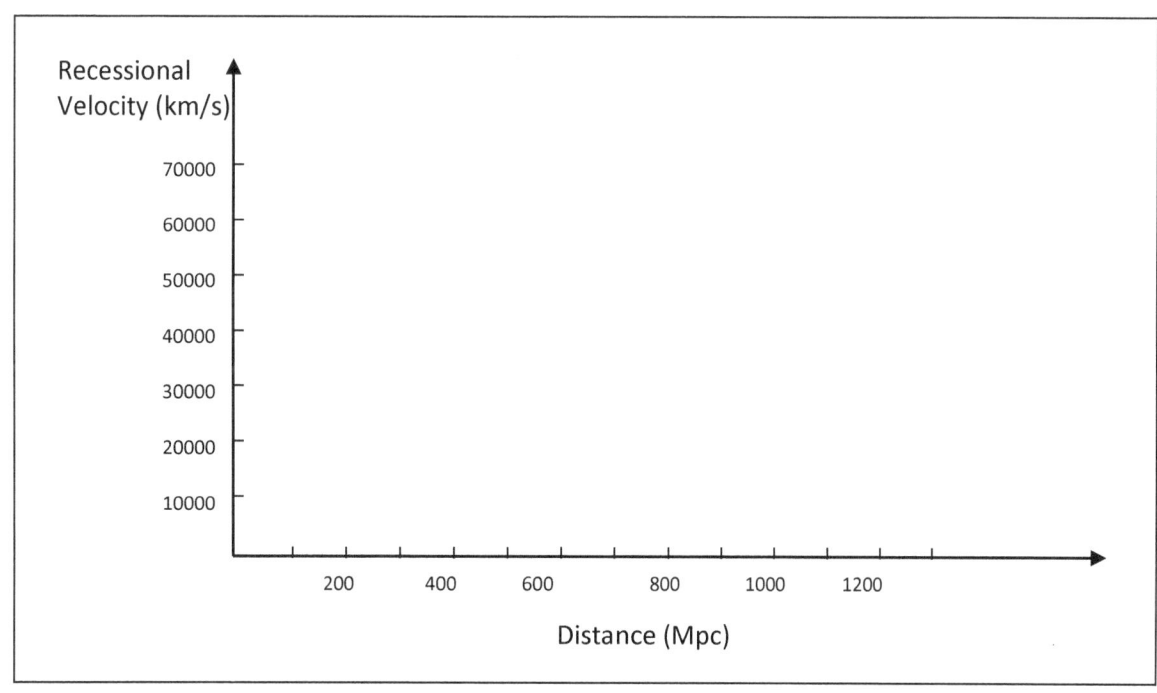

7. Summarize the relationship between recessional velocity and distance.

8. By measuring the red shift in the spectral lines from a newly discovered galaxy it is determined that the galaxy is moving away from us at 30,675 km/s. From your graph, estimate the distance to this new galaxy. Briefly explain your procedure.

Part III: Determining Hubble's Constant

The fact that your data can be reasonably well fit with a straight line means that we can represent the relationship in the simple form

$$v = H \times d$$

where H is called the Hubble constant and is usually expressed in units of km/s/Mpc. That is, H tells us the speed, in km/s, with which a galaxy at 1 Mpc is receding. A galaxy at 1000 Mpc would then be receding at 1000H, and so on.

9. From your graph, estimate the recessional velocity of a galaxy at a distance of 1000 Mpc.

10. What value does this give you for H (just divide by 1000)?

11. The Hubble constant, H, is a measure of the rate at which the universe has been expanding since the big bang to get to its present size. If H had been larger, would the universe have taken more or less time to reach its present size? Explain your answer.

12. Current estimates for H range from 60 to 75 km/s/Mpc. Which end of this range, 60 or 75, predicts the older universe. Explain your answer.

Distances to the Moon and Sun

Please print your name and sign next to it (only those present).

Leader: (C)_____   _____

Explorer: (D)_____   _____

Skeptic: (A)_____   _____

Recorder: (B)_____   _____

Learning Objectives
10. Understand how Aristarchus predicted the relative sizes of the Earth and Moon.
11. Understand how Aristarchus predicted the relative distances to the Moon and Sun.
12. Apply Aristarchus' method to make predictions of relative Moon-Sun distances and Earth-Sun sizes.

Introduction: Following Aristotle (c. 384-322 BC), the most prominent astronomer of his time was Aristarchus (c. 310-230 BC), who was born on the Mediterranean island of Samos. Aristarchus estimated how many times larger the Earth was than the Moon using Earth's shadow on the Moon during a lunar eclipse as illustrated in the figure at right.

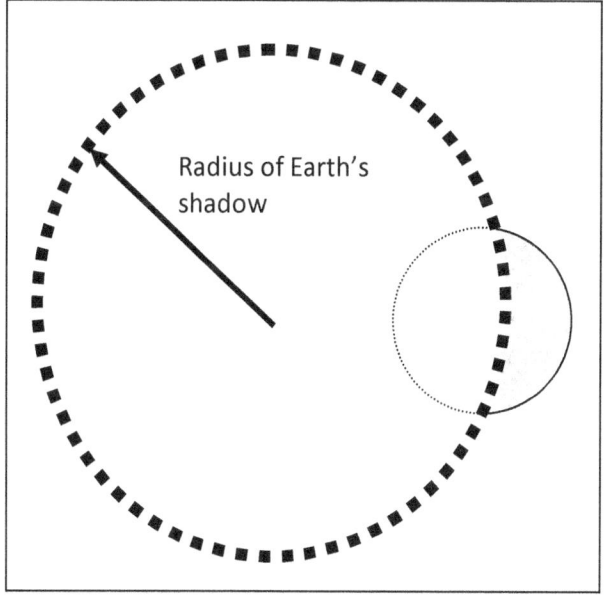

Part I: Relative Sizes

Imagine that Aristarchus found that the Sun was 12 times farther away from Earth than the Moon and that the Earth was 4 times larger across than the Moon.

1. Using the fact that the Moon and the Sun appear about the size in the sky, how many times larger would Aristarchus have concluded the Sun is compared to the Moon (hint: see figure at bottom of page 61)? Explain.

2. How many times larger would Aristarchus have concluded the Sun is compared to the Earth? Explain.

Part II: Size of Earth and Moon

3. Imagine that you were to observe a lunar eclipse and that you made the sketch below of the Moon's appearance part way through the event. This is what you see when you look up at the Moon from Earth. Use this sketch to estimate how many times greater the Earth's diameter is than the Moon's. Remember that the Earth's shadow is a complete circle but you only see it where is covers the Moon.

Explanation of Procedure

Part III: Distance to the Moon and Sun

Aristarchus approach to determining how many times farther away the Sun is than the Moon—called the relative distance—was very prone to error owing to the great distances involved. His method was based on the observation that, when the Moon appears half lit (first quarter phase), the Earth-Moon-Sun angle must be 90^0 as shown in the figure below. So, when the Moon appeared at first quarter, Aristarchus made measurements of the angle between the Moon and the Sun and used simple geometry to determine the relative distances to the Sun and the Moon as shown.

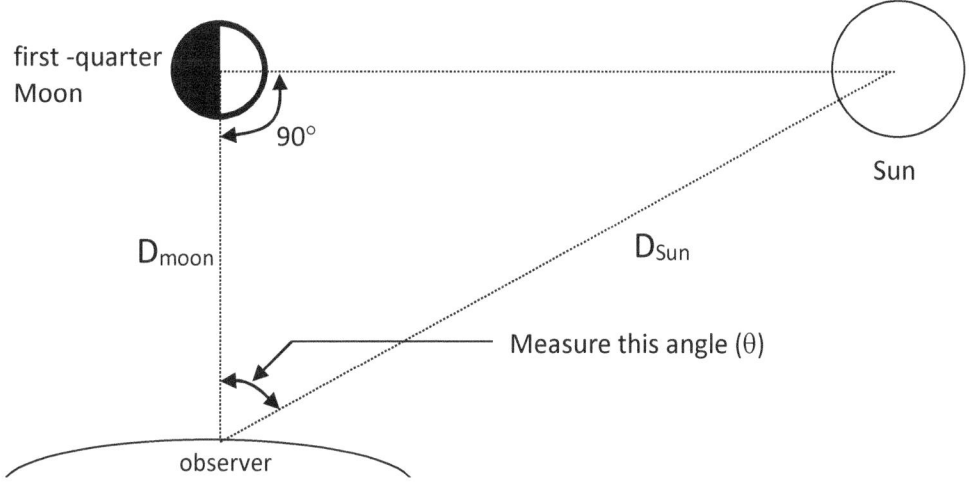

Whoever made this figure was not very careful. The angle shown (marked ☐☐ is only about 60°, which is much too small.

4. Measure the lengths of the lines labeled D_{Moon} and D_{Sun} on the figure above and estimate how many times farther away the Sun is than the Moon (i.e., what is D_{Sun} divided by D_{Moon} ?).

5. We now know that D_{Sun} is many more times D_{Moon} than in the above figure. Does this mean that the real value of θ is greater or less than 60°? Explain.

6. Imagine that you were to repeat Aristarchus' measurement and found the angle θ to be 81°. Your task is to use a scale diagram (found on last page) and follow the procedure on the next page to determine the relative distance of the Sun compared to the Moon (D_{Sun} divided by D_{Moon}).

> Steps:
> 1. Measure 2.0 cm along the vertical line extending straight up from the observer and place an **X** to indicate the position of the Moon in your scale diagram (D_{Moon} = 2.0 cm).
> 2. Draw a horizontal line passing through the Moon's position and extending to the right edge of the page. Because the Moon is exactly half lit, the Sun necessarily lies along this line.
> 3. Draw a line at the angle θ from the observer's position. Where this line intercepts the horizontal line must represent the position of the Sun.
> 4. Measure the distance to the Sun (D_{Sun}) and enter it in the box.
> 5. Estimate how many times farther away the Sun is than the Moon and enter it in the box.

7. Use your result from above to estimate how many times larger across the Sun is than the Moon. Explain.

8. Use your result from Part II as well as question 8 above to determine how many times larger across the Sun is than the Earth. Explain your reasoning. (Hint: you can review the procedure in Part I.)

9. Aristarchus concluded that the Sun is about 7 times farther across than the Earth—we now know the value to be about 109. Which of the two measurements, the size of the Moon compared to Earth or the distance to the Sun compared to the Moon, was responsible for the large error in his conclusion. Explain.

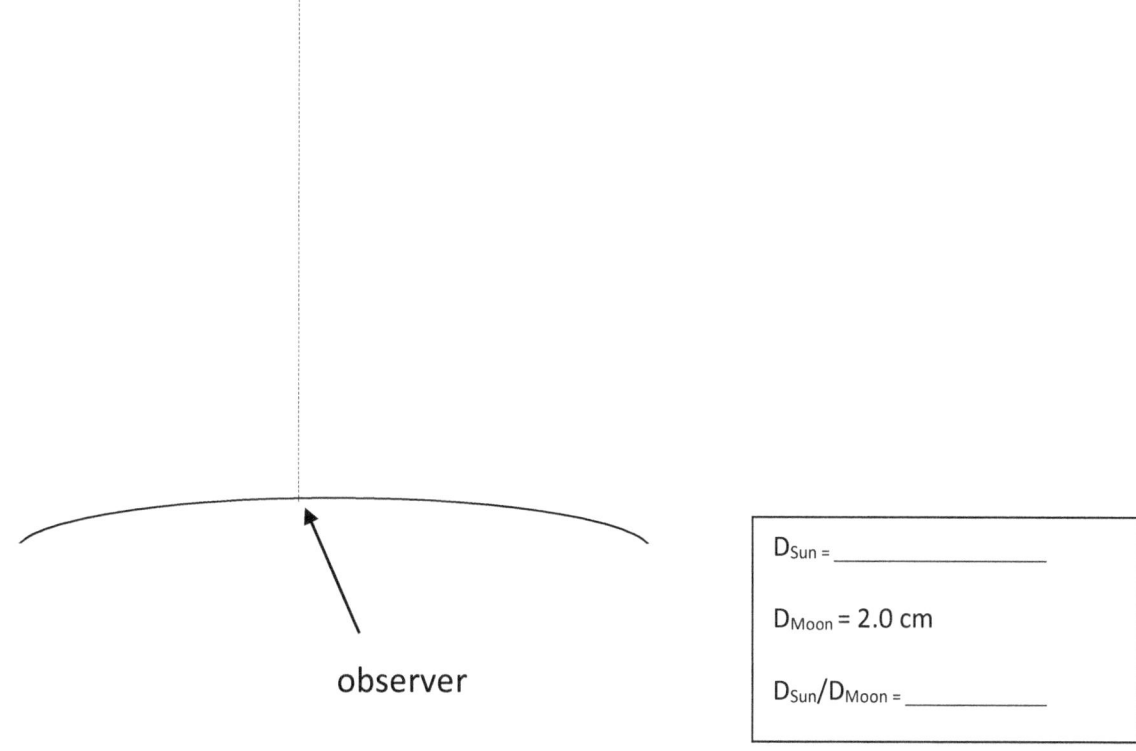

D_{Sun} = _____	
D_{Moon} = 2.0 cm	
D_{Sun}/D_{Moon} = _____	

Question: Aristarchus got a value for D_{Sun}/D_{Moon} of 20. Was the angle he measured greater or less than 81^0? Explain your answer.

Tracing Epicycles

Please print your name and sign next to it (only those present).

Leader: (A)_____   _____

Explorer: (B)_____   _____

Skeptic: (C)_____   _____

Recorder: (D)_____   _____

Learning Objectives
13. Given a deferent and epicycle, trace Ptolemaic planetary orbits as seen from the North Star.
14. Describe how the epicycle model accounts for the observed retrograde motion of planets.

Part I:

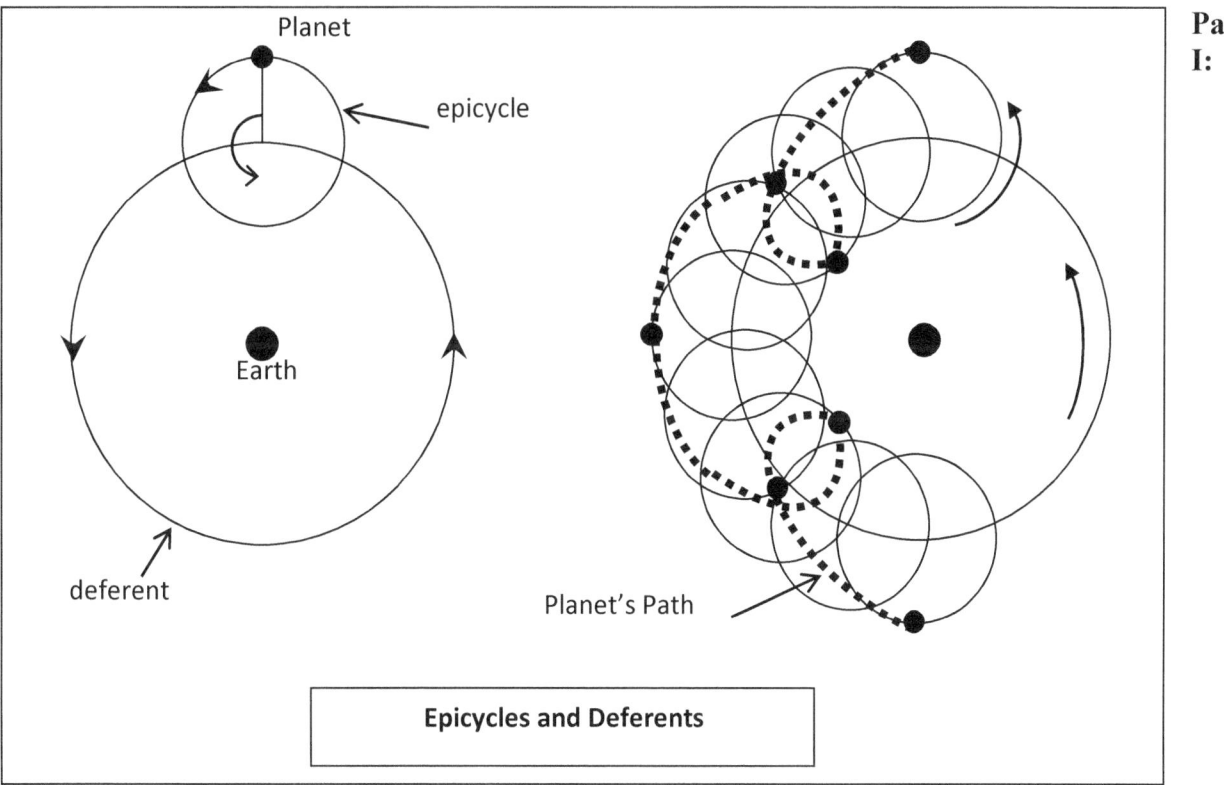

Epicycles and Deferents

Observing Retrograde Motion

The most difficult part of constructing an accurate model for planetary motions is that planets seem to wander among the stars. Normally, planets move from west to east as seen against the background stars, but occasionally (and predictably) they stop and reverse direction for several weeks. This backwards motion is called retrograde motion.

1. Given the data in table 1, plot the motion of a planet on the chart below (record dates next to each position plotted). Then, connect a smooth line showing the path of the planet through the sky.

197

Figure 1 – Planet Path

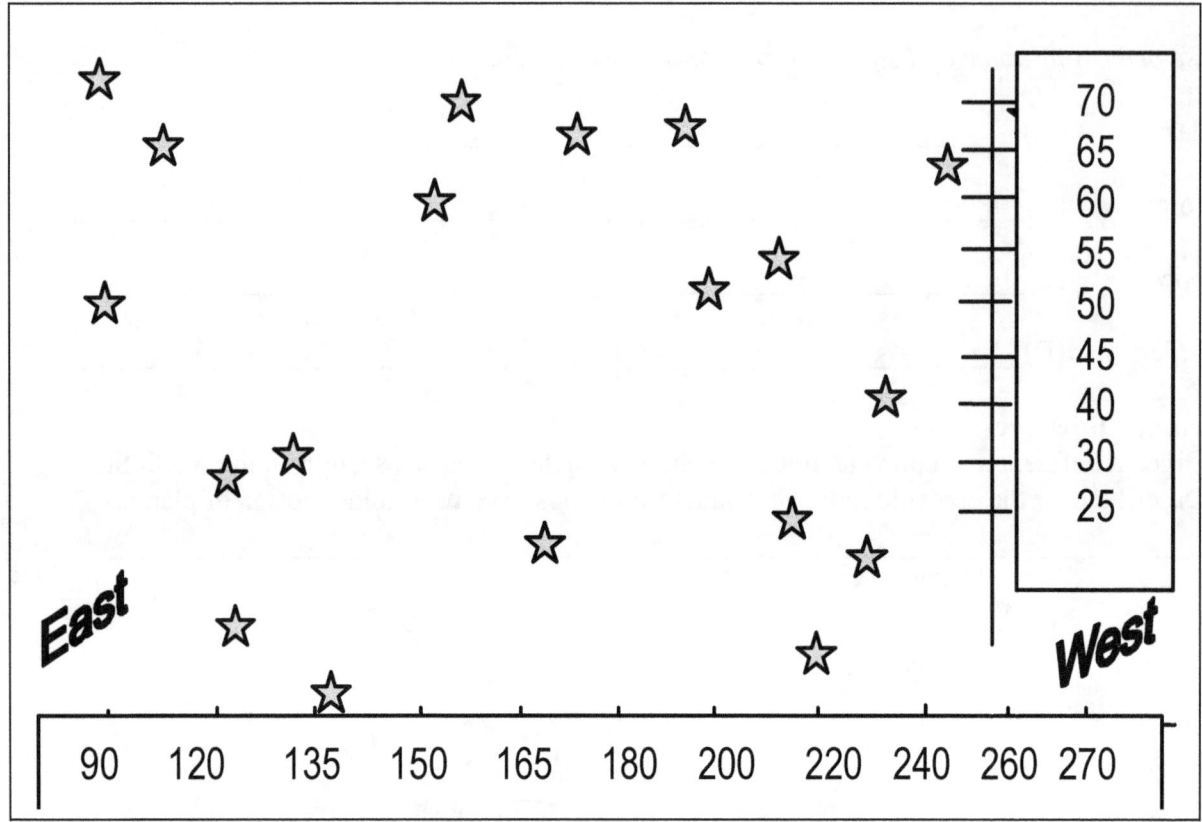

Table 1 – Mystery Planet Positions

Date of Observation	Azimuth Direction (horizontal)	Altitude Direction (vertical)
May 1	240	45
May 15	210	55
June 1	170	50
June 15	150	40
July 1	170	30
July 15	180	45
August 1	140	50
August 15	120	52

2. Approximately how many days does this mystery planet spend going retrograde? How did you determine this number?

Part II: Tracing Orbits

Shown at right is the orbit of an imaginary planet orbiting Earth as conceived in the Ptolemaic Geocentric System. The planet's epicycle has an orbital period equal to five times the orbital period of the deferent (5 ep. = 1 def.). Your task is to sketch the planetary orbits that would correspond to:
(a) 2 ep = 1 def. *and*
(b) 4 ep = 1 def.

1 def. = 5 ep.

Be sure everyone in the group can do this task.

Planet's Path

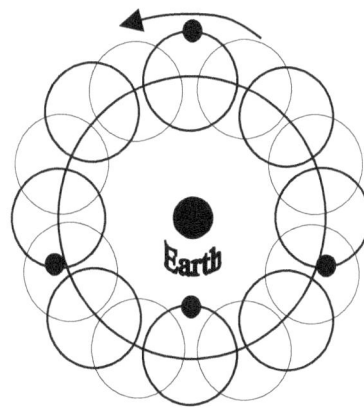

1 def. = 2 ep.

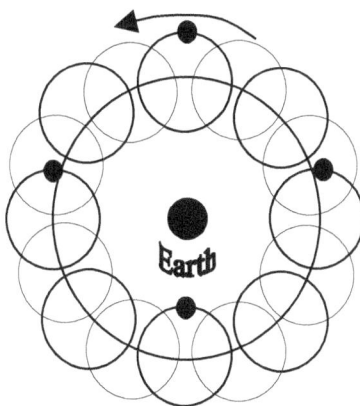

1 def. = 4 ep.

3. Suppose your instructor says that Mars is in retrograde motion tonight and will rise at midnight. In what direction would you look to see it? Explain.

Determining the Orbit of Mars

Please print your name and sign next to it (only those present).

Leader: (B)_____   _____

Explorer: (C)_____   _____

Skeptic: (D)_____   _____

Recorder: (A)_____   _____

Learning Objectives
15. Use Kepler's method to determine Mars' orbit from data listing the bearings to Mars taken at one-Martian-year intervals.
16. Use the map of Mars' orbit to determine the radius of the orbit in astronomical units (AU).

Introduction: Johannes Kepler (1571-1630) had the goal of developing an accurate mathematical description of the motions of the planets, a goal that he did eventually achieve. However, first he needed to determine exactly what those motions were and, unlike most astronomers before him, Kepler chose to look only at that data to answer this question without relying on any underlying philosophical models. We will look now at how Kepler used Tycho's data to determine the true orbit of Mars.

Procedure: Kepler's approach was very clever. He knew that the Earth goes around the Sun every 365.25 days and was able to work out the correct numbers for the other planets as well. For instance, Mars goes once around the Sun every 687 days, which is about one and seven eighths Earth years. The table at the bottom of the final page gives matched data for eight different locations of Mars around its orbit. Mars is in the same location for both observations (taken 687 days apart) but Earth is in different locations. Your task is to use this data to plot Mars' path around the Sun by finding these eight locations on the scale diagram. Mars position 3 has been done for you using the following procedure.

Suppose that we observe Mars on January 1 at which time the Earth is at position **C** in the figure on the final page. Mars is observed at an angle of 10° using the coordinate system on the figure. If you draw a straight line beginning at position **C** at 10°, Mars must lie somewhere along that line. Mars is next observed 687 days later, in mid-November, when Mars has returned to the same location but Earth is now at position **B**. From Earth, Mars now appears to lie at an angle of 38°. If you draw a straight line beginning at position **B** at 53°, Mars must lie somewhere along that line. The intersection of your two lines pinpoints Mars location on both dates and gives you one point in its orbit.

Use the remaining data to locate seven additional points along Mars' orbit making sure that each member of the group finds at least one point (this should take a total of 10-15 minutes). When you have all the points, sketch in the approximate path of Mars during this time interval—you will only have about half an orbit.

1. From your map, estimate the radius of Mars' orbit in Astronomical Units. (*Hint: Measure the approximate radius of Mars' orbit and divide that by the approximate radius of Earth's orbit*).

2. Which one of the eight points in Mars' orbit was most prone to error? Why?

3. In the orbit you have drawn, there is no epicycle evident. Why not?

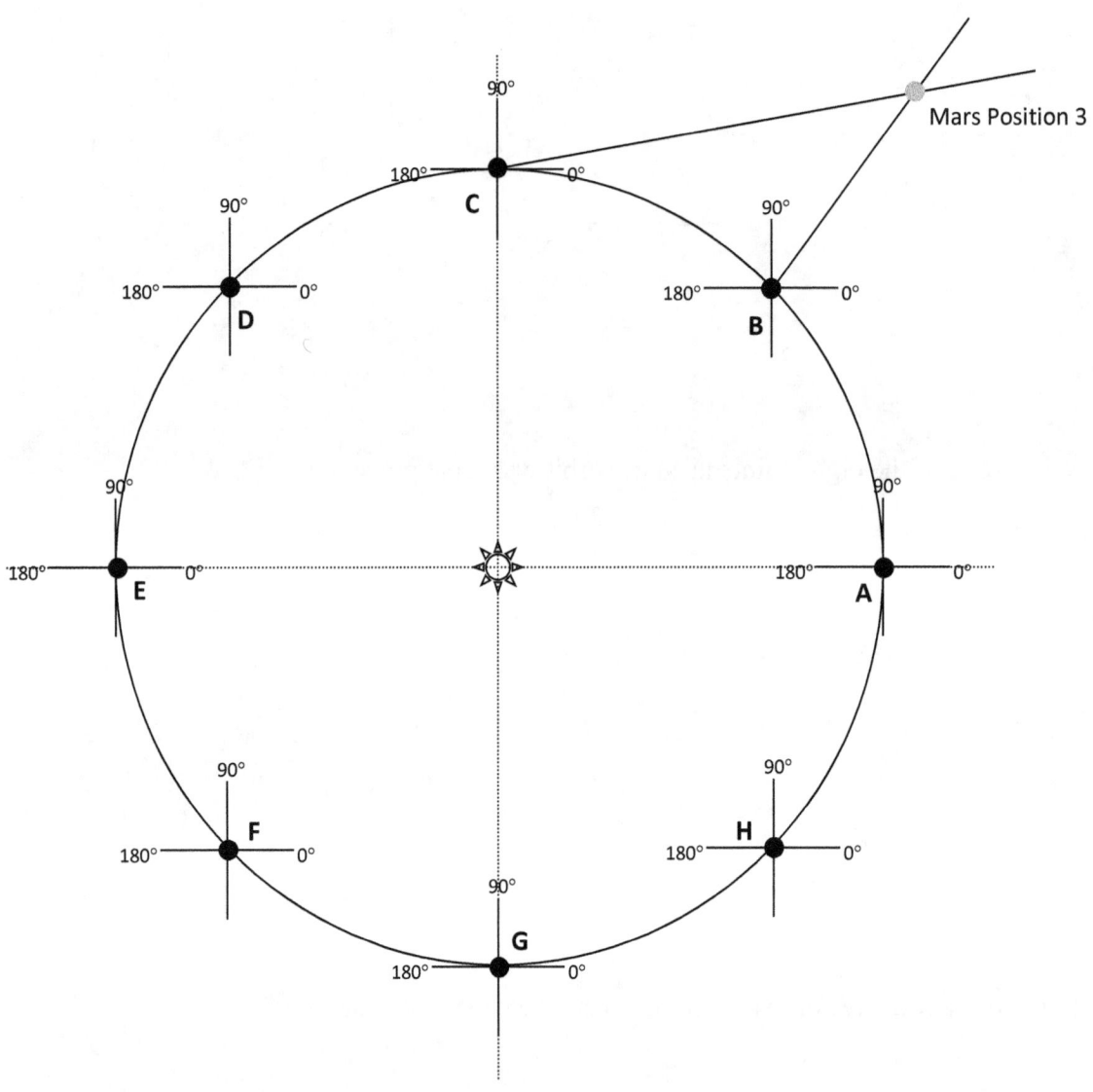

Mars Position	Earth's Position	Bearing to Mars	Earth's Position One Martian Year Later	Bearing to Mars
1	A	0	H	38
2	B	-4 (or +356)	A	55
3	C	10	B	53
4	D	34	C	46
5	E	62	D	58
6	F	93	E	82
7	G	124	F	109
8	H	155	G	139

The Phases of Venus

Please print your name and sign next to it (only those present).

Leader: (C)_____   _____

Explorer: (D)_____   _____

Skeptic: (A)_____   _____

Recorder: (B)_____   _____

Learning Objectives
17. Use the Ptolemaic model to predict the phases of the Venus.
18. Use the Copernican model to predict the phases of the Venus.
19. Understand how Galileo's observation of the phases of Venus helps to select between these two models.

Introduction: As the first scientist to undertake detailed astronomical study with the aid of a telescope, Galileo Galilei (1564-1642) was able to make a number of important discoveries that greatly influenced our understanding of the universe. These observations were published in 1610 in his book *The Starry Messenger*, which was written in a style that made it accessible to a wide audience. One of his observations was that Venus exhibits phases much like our Moon.

Part I: Venus in the Ptolemaic Model

In the Ptolemaic model, Venus moves on an epicycle whose center is always directly between Earth and the Sun. The figure on page 211 shows Venus in a series of locations around its epicycle. Your task is to <u>sketch</u> the appearance of Venus, as observed through a telescope from Earth, when Venus is at each of the five indicated positions using the following procedure.

- For each of the five positions <u>along the epicycle</u> (i.e., the top half of the page) shade in the dark half of Venus.
- For each of these positions sketch, in the spaces at the bottom of the page, what Venus would look like as seen from Earth. In your sketches, shade the portion that would appear dark and the leave alone the part that would appear bright.

1. Is there any location along the epicycle at which Venus would appear as a near fully lit disk? (If there is, indicate which one.)

Part II: Venus in the Copernican Model

In the Copernican model, Venus moves around the Sun in an orbit that is only 72% the size of Earth's orbit. The figure on page 213 shows Venus at five different locations around the Sun. Your task is to repeat the procedure from Part I to sketch the appearance of Venus as seen from Earth when Venus is observed at each of these locations.

2. Is there any location along its orbit at which Venus would appear as a near fully lit disk? (If there is, indicate which one.)

3. Galileo observed that Venus goes through a complete set of phases including full phase. With which system of planetary motion, Ptolemaic or Copernican, is this observation consistent? Explain.

4. Do these observations of the phases of Venus necessarily confirm the heliocentric model? (Hint: think about Tycho's model in which the Sun orbits the Earth but Venus orbits the Sun.) Explain your answer.

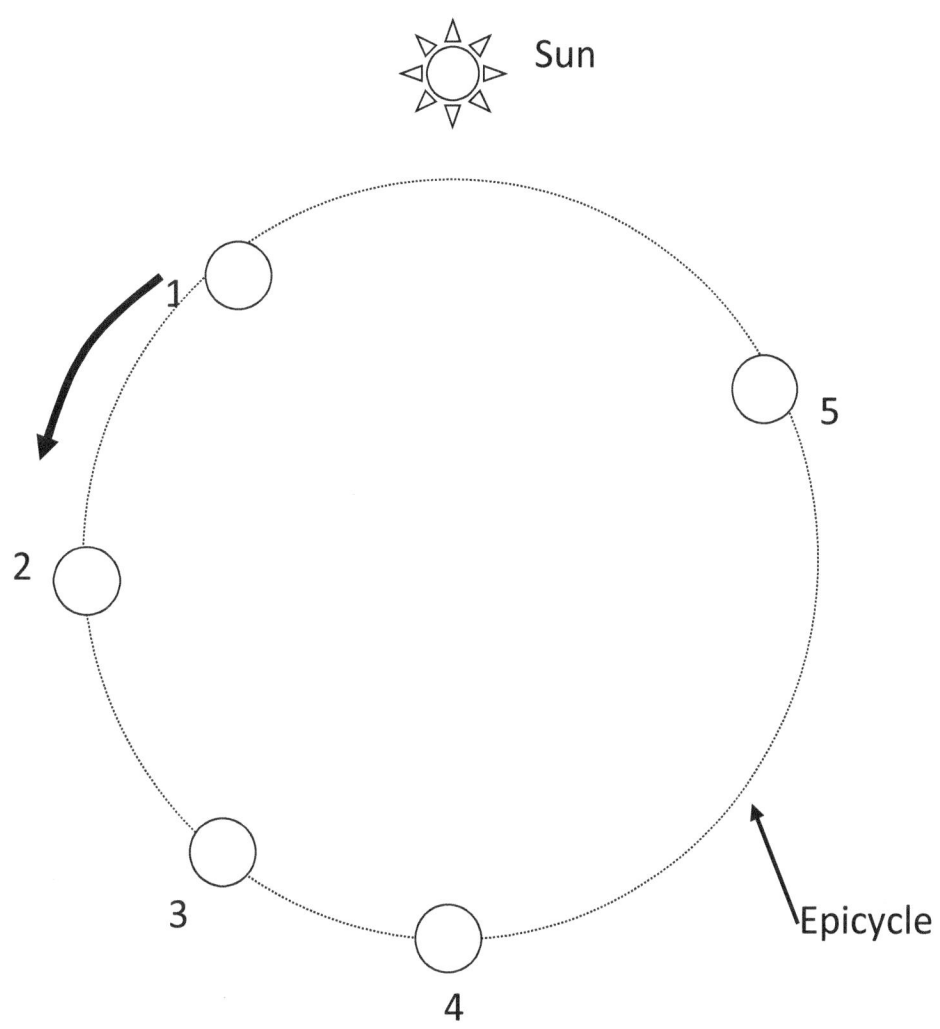

Sketch Venus's appearance as seen from Earth at the five locations shown above.

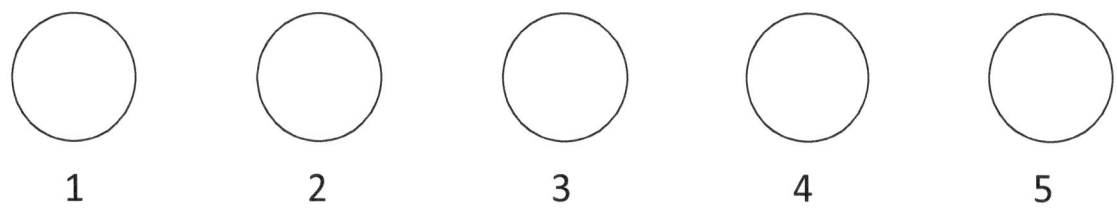

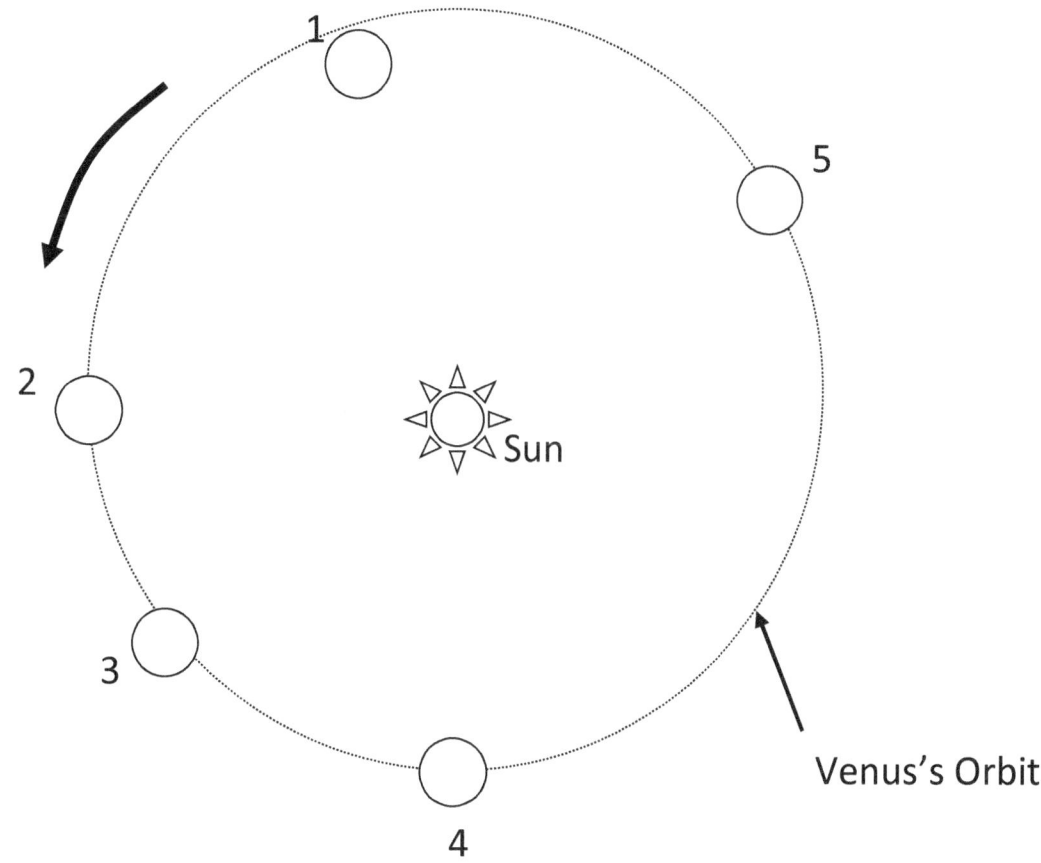

 Earth

Sketch Venus's appearance as seen from Earth at the five locations shown above.

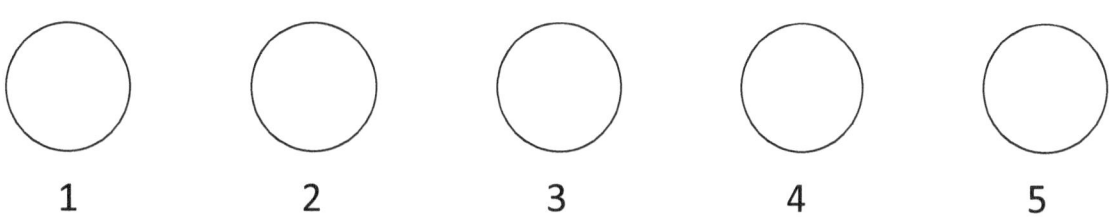

Mapping the Solar System From Earth

Please print your name and sign next to it (only those present).

Leader: (D)_____   _____

Explorer: (A)_____   _____

Skeptic: (B)_____   _____

Recorder: (C)_____   _____

Learning Objectives
1. Comprehend that the observer's position on Earth makes particular objects in the sky visible at specific times.
2. Analyze the rotation of an Earth observer to predict the rising & setting times of sky objects.
3. Synthesize heliocentric object locations and interpret to a geocentric perspective.
4. Synthesize geocentric object positions and interpret to a heliocentric perspective.

Background: Some newspapers and science magazines, such as *Sky and Telescope*, provide sky charts that describe what sky objects are visible at different times. These typically include prominent stars, bright planets, and the Moon. There are two principle maps provided to readers: (1) a geocentric horizon view and (2) a heliocentric orrery view. The *geocentric* perspective is the view from Earth looking up into the southern sky. The *heliocentric* perspective is the view of the Solar System looking down from above. From above, the plants orbit and spin counter-clockwise (except Venus, which appears to spin backwards).

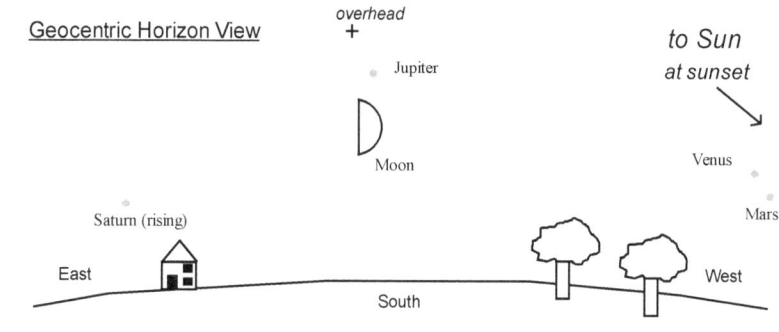

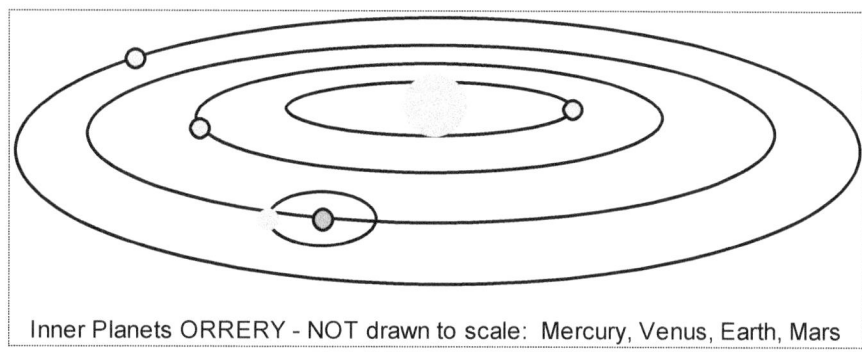

Part I: Rising and Setting Times

As seen from above, Earth appears to rotate counterclockwise. *Figure I-a* shows a top view of Earth and an observer at noon. Note that our Sun appears overhead when standing at the equator.

Figure I-a: Observer Positions on Earth
[Observer is at Equator]

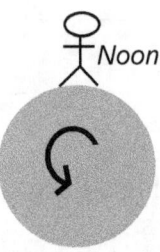

1. In Figure I-a, sketch and label the positions of the observer at midnight, 6 pm (sunset) and 6 am (sunrise).

2. Consider *Figure I-b,* which shows Earth, Moon, Mars, and Venus. At what time would each of these sky objects be overhead? Remember that Earth spins counter-clockwise when viewed from above. [*Hint: Make use of Figure I-a*]

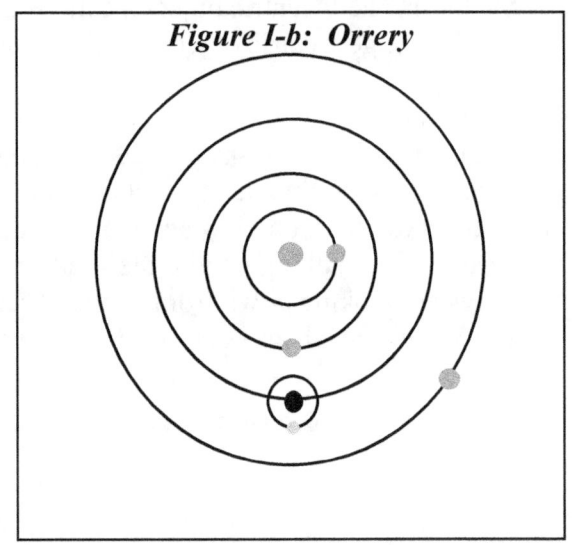

Time Overhead:

Venus: _____

Moon: _____

Mars: _____

3. If Earth spins in 24 hours, that means that each sky object is visible for about 12 hours. What time will the sky objects shown in Figure I-b rise and set? Complete the table below? *Each member of your team should fill in the data for one sky object.*

Sky Object	Rise Time	Time Overhead	Set Time
Sun			
Venus			
Moon			
Mars			

4. Using complete sentences, explain why our Sun is not visible at midnight. Add a sketch of Earth, Sun, and observer in the space provided to support your explanation.

Narrative	*Sketch*

Part II: Converting Geocentric to Heliocentric

5. *Figure II-a* shows the horizon view of the first quarter Moon and Saturn visible at sunset. On the orrery shown in *Figure II-b*, sketch and label the position of Jupiter, Moon and Saturn. Use an arrow to indicate the direction to our Sun. Start by indicating the position of the observer at sunset. After completing the diagram, complete the table.

Sky Object	Rise Time	Set Time
Sun		
Jupiter		
Moon		
Saturn		

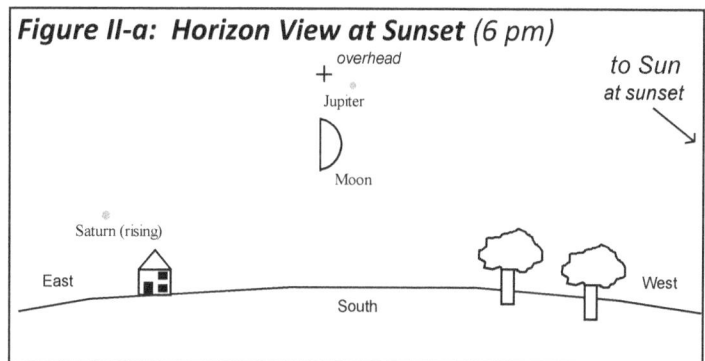

Figure II-a: Horizon View at Sunset (6 pm)

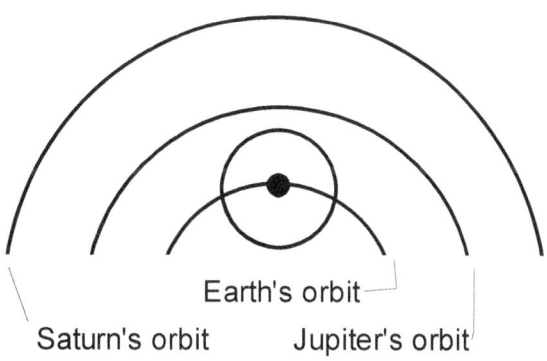

Orrery Not Drawn to Scale !!
Figure II-b

6. If Neptune is visible overhead in the southern sky at sunrise (6 am) sketch the relative positions of Sun, Earth, Neptune, and observer in an orrery in the space below.

Part III: Converting Heliocentric to Geocentric

7. *Figure III-a* shows the position of Mercury, Venus, Earth, Mars and Moon. On the horizon diagram, *Figure III-b*, sketch and label the positions of Mercury, Venus, Mars, a comet, and Moon at <u>midnight</u>.

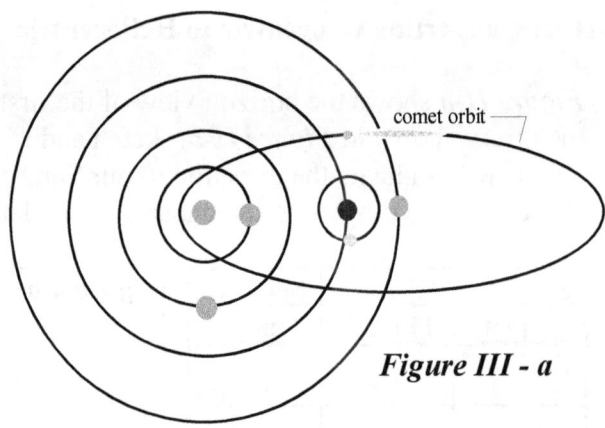

Figure III - a

Figure III-b: Geocentric Horizon View at Midnight

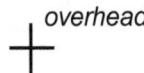

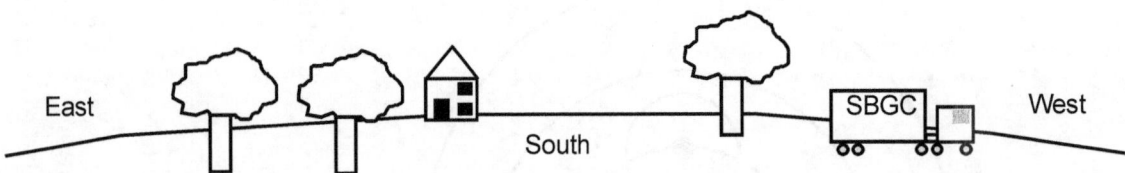

8. Venus is often called the *morning star* or the *evening star*. Why is it never seen at midnight?

Part IV: Current Events

9. Your instructor will provide you with a magazine photocopy or software print-out of the current sky or orrery. Convert the given current sky to an orrery OR convert the current orrery to a horizon view (you can select the time). Make your sketches in the space below. Be certain to label every item carefully.

10. Describe and sketch the night sky if you were to go outside at midnight tonight (*assume there are no clouds*).

Moon Phases

Please print your name and sign next to it (only those present).

Leader: (A)_____   _____

Explorer: (B)_____   _____

Skeptic: (C)_____   _____

Recorder: (D)_____   _____

Learning Objectives
1. Develop a mental model of the Sun-Earth-Moon geometry responsible for lunar phases.
2. Predict the phase of the Moon given the relative positions of the Sun-Earth-Moon system.
3. Predict the relative positions of the Sun-Earth-Moon system given the Moon phase and time of day.

Introduction: Throughout history, many cultures have used the lunar cycle as a method to measure increments of time—a lunar calendar. Nomadic peoples regulated their calendar based entirely on the 28-day cycle of the Moon. Every time the slender crescent after a new Moon appeared in the western evening sky, a new month began at the evening hour. The lunar cycle was quite practical because of its short duration and its ease in use. The phases of the Moon are due to the changing relative positions of the Sun, Moon, and Earth. One half of the Moon is always lit by the Sun. As the Moon orbits around the Earth, we see first more and more of the lit half and then, less and less.

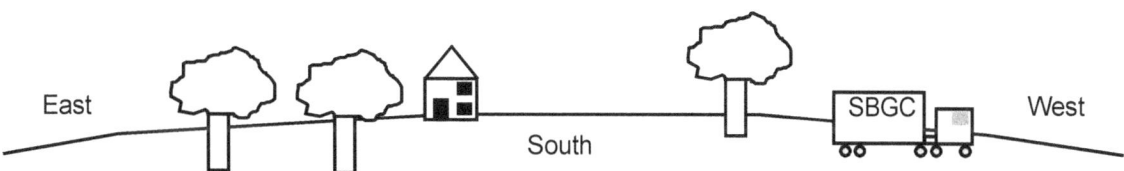

For all of these Moon phase activities, assume that you are facing south. When the Moon is setting in the west it will be directly to your right. When it is rising in the east it will be directly to your left. When the Moon is straight up from the southern horizon—and fairly high overhead—we will refer to it as being south.

Part I:

The figure below shows an observer on Earth's equator as seen from above the north pole. The Earth appears to spin counterclockwise when viewed from above.

1. Draw and label the observer where she would be at midnight, sunset (6 PM), and sunrise (6 AM). For the 6 AM and 6 PM locations, draw and label arrows to indicate east and west.

Part II:

Because in our everyday experience we tend to think about the Moon and Sun moving about a stationary Earth, we will now imagine looking down upon our observer from the above the north pole and spinning with the Earth. The figure below shows the direction from which the sunlight would be coming at noon for the observer positioned as shown.

2. Draw arrows showing the direction from which the <u>sunlight would be coming</u> at sunset, midnight, and sunrise at the observer's location. Label your arrows accordingly. (Hint: if you are stuck, refer back to Part I.) Draw and label arrows to indicate east and west at the observer's location.

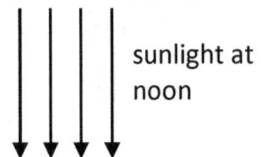

Part III:

The figure below shows the Earth-Moon-Sun system as viewed from above. The Sun always illuminates the half of the Moon facing the Sun. It looks like the Earth might block the sunlight from reaching the Moon but this rarely happens because the Moon is normally above or below a straight line drawn from the Sun to the Earth as shown at right.

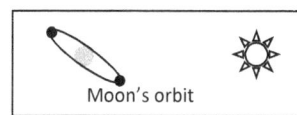
Moon's orbit

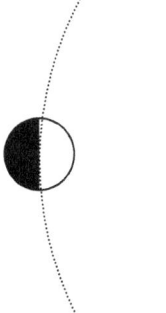  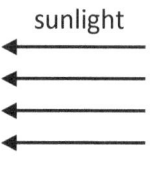
sunlight

3. What time is it at the observer's location in the figure above?

4. The observer looks to the west to see the Sun. What direction does she look to see the Moon?

5. In the box at right, sketch what the observer would see looking at the Moon (shade in any portion that would not be visible). What do we call this phase (circle)? The last page of this activity is a *Moon Phase Poster*, giving the names of the different phases.

    new   crescent   quarter   gibbous   full

    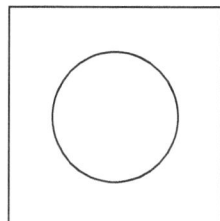

6. At midnight, the Sun will be on the opposite side of the Earth from the observer and the Moon will be overhead. What phase will the Moon appear then (circle)?

    new   crescent   quarter   gibbous   full

7. In what direction would the observer look to see the Moon six hours later at sunrise?

8. What phase will the Moon appear at sunrise (circle)?

    new   crescent   quarter   gibbous   full

9. Will the Moon be visible at noon this same day?
   Explain your answer with a sketch at right.

Part IV:

The Moon is in its new phase when the lit side is on the far side of the Moon from the observer. The figure below shows the observer at sunset with the Sun in the west.

10. Sketch the Moon on the figure above so that the observer would say it is in its new phase. As was done in Part III, shade in the dark side of the Moon to show which side of the Moon is actually illuminated by the Sun.

11. At what time will the Moon set below the horizon?

12. At what time will the Moon rise <u>and</u> in what direction would you look to "see" it (assuming it could actually be seen)?

Part V:

13. A few days later, the Moon will be about half way between overhead and the western horizon as the Sun is setting in the west as shown on the figure below. Again, the side illuminated by the Sun is shown. Remember that the Sun is very far away. In the box at right, sketch what the observer would see looking at the Moon (shade in any portion that would not be visible). What do we call this phase (circle)?

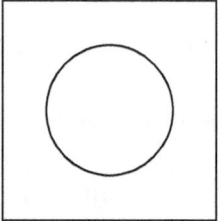

new   crescent   quarter   gibbous   full

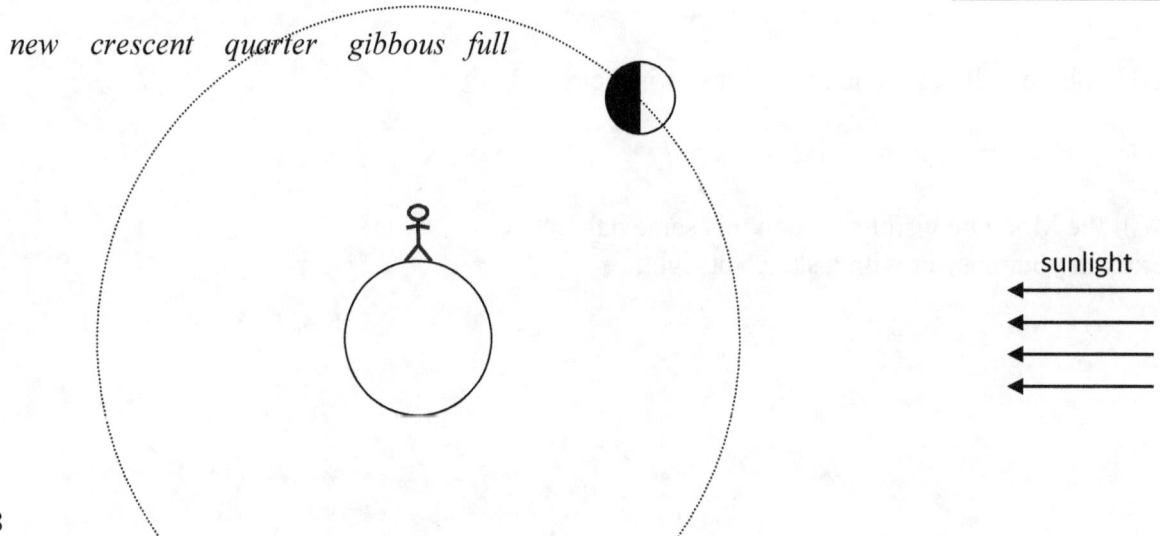

218

14. At about what time will this Moon set?

15. At about what time will this Moon rise?

16. In the box at right, locate the Moon and the direction of the sunlight as this Moon is rising.

17. In the Northern Hemisphere, we often say that the Sun is in the south when at its highest point, which we have called overhead until now. We can then describe the Sun's position changing throughout the day with the sequence east (6 AM), southeast (9 AM), south (noon), southwest (3 PM), and west (6 PM). In what direction does the observer look to see the Sun when the crescent Moon is rising (circle)?

east   southeast   south   southwest   west

18. In what direction would the observer look to see the crescent Moon at noon (circle)?

east   southeast   south   southwest   west

19. If the observer were looking toward the Moon at noon, would she see the Moon's illuminated crescent on the right side or the left side (circle)?   *right   left*

Part VI:

About one week after the new Moon, the Moon will be to the south (overhead) at sunset.

20. In the box at right, locate the Moon and the direction of the sunlight at this time. Shade in the dark side of the Moon to show which side of the Moon is actually illuminated by the Sun.

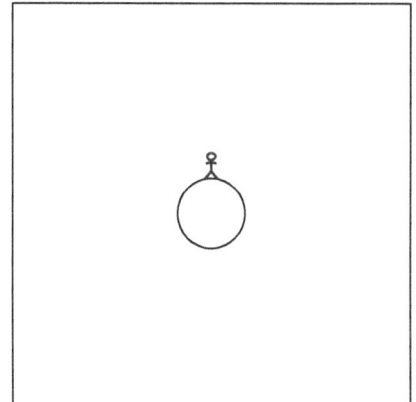

21. In the box at right, sketch what the observer would see looking at the Moon (shade in any portion that would not be visible). What do we call this phase (circle)?

new   crescent   quarter   gibbous   full

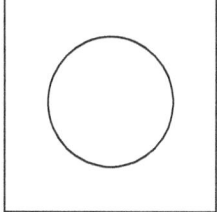

22. Because the Moon can take on this same basic appearance (half lit, half dark), we differentiate between a first quarter Moon, which occurs about one week after the new Moon and a third quarter Moon, which occurs about three weeks after the new Moon. If the observer were looking toward a first quarter Moon at sunset, would she see the Moon's illuminated portion on the right side or the left side (circle)?

   right   left

23. In the box at right, indicate and label the locations of the first quarter Moon at each of the times below. In cases where the Moon is above the horizon, describe its direction as east, southeast, south, southwest, or west. If it is below the horizon, state that explicitly. The first one is done for you.

   6 PM: *south*

   9 PM:

   Midnight:

   6 AM:

   Noon:

   3 PM:

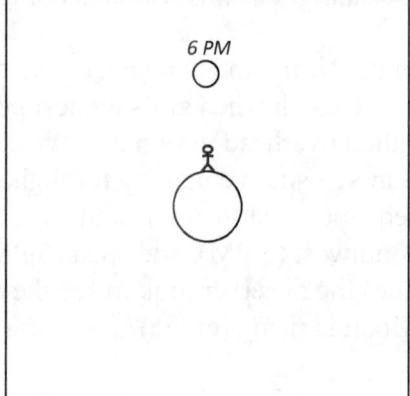

Part VII:

About a week and a half after a new Moon, the Moon will be to the southeast (half way between the eastern horizon and overhead) at sunset.

24. In the box at right, locate the Moon and the direction of the sunlight at this time. Shade in the dark side of the Moon to show which side of the Moon is actually illuminated by the Sun.

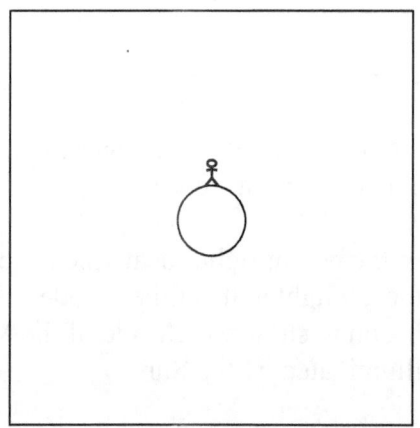

25. In the box at right, sketch what the observer would see looking at the Moon (shade in any portion that would not be visible). What do we call this phase (circle)?

   new   crescent   quarter   gibbous   full

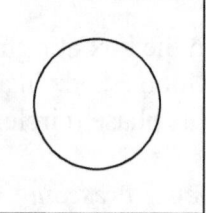

26. Any time the Moon's lit portion is growing from one night to the next we say that it is **waxing**. When the Moon's lit portion is shrinking we say that it is **waning**. Is the Moon in this part waxing or waning? Explain how you know.

27. Imagine that you see a waxing crescent Moon setting on the western horizon. In the box at right, locate the Moon and the direction of the sunlight at this time. Shade in the dark side of the Moon to show which side of the Moon is actually illuminated by the Sun. About what time is it?

Part VIII:

The third quarter Moon occurs about 3 weeks after new Moon. Whereas a first quarter Moon appears lit on the right side, a third quarter Moon appears lit on the left side when in the south (overhead).

28. In the box at right, locate the Moon and the direction of the sunlight when the Moon is overhead. Shade in the dark side of the Moon to show which side of the Moon is actually illuminated by the Sun. About what time is it?

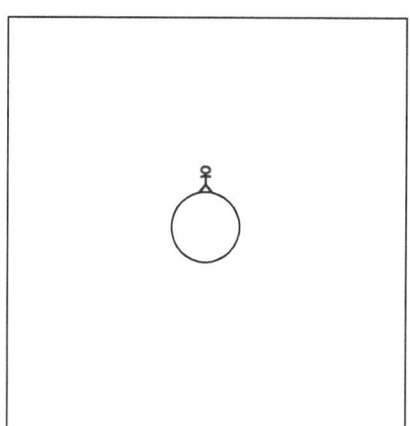

29. At what time does the third quarter Moon rise? Explain your reasoning.

30. At what time does the third quarter Moon set? Explain your reasoning.

31. You see a third quarter Moon in the southwest sky. In the box at right, locate the Moon and the direction of the sunlight at this time. Shade in the dark side of the Moon to show which side of the Moon is actually illuminated by the Sun. About what time is it?

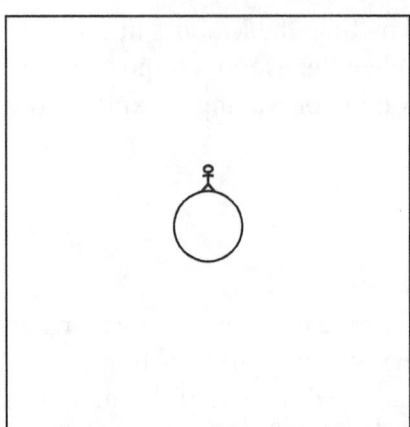

Part IX: Putting it all together

Work as a group to find the correct answer to each of the following.

32. Which one of the following is a possible observation at 4 a.m.?
    a. a full Moon just rising in the east
    b. a crescent Moon just setting in the west
    c. a crescent Moon rising in the east
    d. a gibbous Moon rising in the east
    e. a first-quarter Moon in the south-west sky

33. An astronaut is standing on the Moon when the Moon is in a crescent phase. In what phase does the Earth appear to the astronaut? It will help to draw a diagram.
    a. full
    b. gibbous
    c. quarter
    d. crescent
    e. new

Moon Phase Poster

	New Moon	Day 0 and Day 28
	Waxing Crescent	Days 1-6
	First Quarter	Day 7
	Waxing Gibbous	Days 8-13
	Full Moon	Day 14
	Waning Gibbous	Days 15-20
	Third Quarter	Day 21
	Waning Crescent	Days 22-27

www.ingramcontent.com/pod-product-compliance
Lightning Source LLC
Chambersburg PA
CBHW080653190526
45169CB00006B/2096